FROM A TO BEE

FROM A TO BEE

Summersdale Publishers Ltd
46 West Street
Chichester
West Sussex
PO19 1RP
UK

www.summersdale.com

Printed and bound by CPI Group (UK) Ltd, Croydon

ISBN: 978-1-84953-272-3

FROM A TO BEE

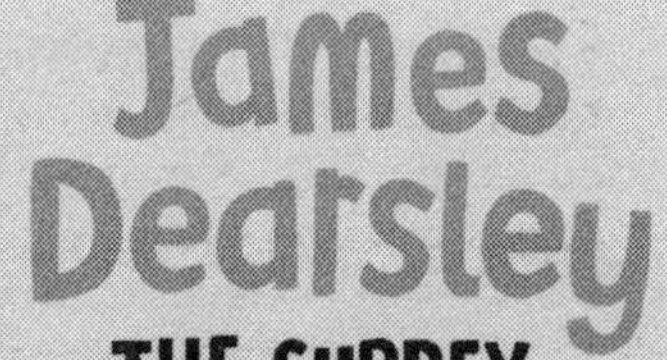

THE SURREY BEEKEEPER

ACKNOWLEDGEMENTS

This book is dedicated to Mum and Dad for their unwavering support over the years and to my sister Emma, to my lovely Belle-Mère and also to Peter who is sorely missed but never far from our thoughts. However, my darling Jo deserves all the credit for putting up with my crazy plans and ideas – for which I am eternally grateful. I am proud to be her husband each and every day. Finally, this book is dedicated to my beautiful boys, Sebastian and Edward, with whom I look forward to a lifetime of adventures and mischief.

I ran a social media competition to name the title of this book and so I must personally thank everyone that suggested a title. The winner, *From A to Bee*, was suggested by Henrik Cullen, but I also have to extend my thanks to my good friend Rob Hoye, who was beaten into second place by a mere seven votes. Another good friend, George TC, came joint third with Liz Bennett. It was great fun and thank you to all that took part and thank you to Summersdale, who allowed me to run this rather madcap campaign and have been supportive throughout and a joy to work with.

ABOUT THE AUTHOR

James Dearsley, the Surrey Beekeeper, started The Beginner Beekeepers page on Facebook, one of the largest online communities of beekeepers, and is on Twitter (@surreybeekeeper). His site **www.surreybeekeeper.co.uk** started as a blog, so others could learn from his mistakes, and expanded into a shop and general online resource for beekeepers. He has written for a variety of publications around the world including *The Ecologist* and has recorded a DVD, *Beekeeping for Beginners*, with Charlie Dimmock, which is now on general release. He lives with his wife and two sons in Surrey.

CONTENTS

THE BEEHIVE

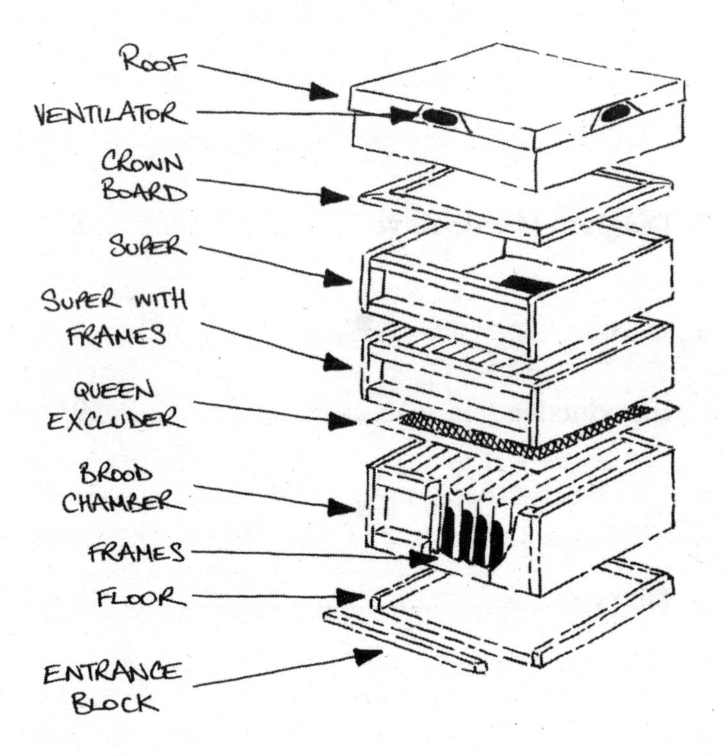
ROOF
VENTILATOR
CROWN BOARD
SUPER
SUPER WITH FRAMES
QUEEN EXCLUDER
BROOD CHAMBER
FRAMES
FLOOR
ENTRANCE BLOCK

INTRODUCTION

Beekeeping… Oh my, what have I done?

I am thirty years old, have been married for three years and am a new father to a fantastic little boy. Surely there are things that I should be doing at this age which do not involve little yellow and black insects that can hurt you if you are remotely clumsy – which, at 6 foot 5 inches, I have an amazing ability to be. My wife, Jo, thinks I have lost my mind, and my little boy looks at me rather strangely when I start running around the living room making buzzing noises and flapping my arms frantically as I try desperately to make him laugh. I think maybe my wife is right. My mother has somewhat disowned me and blames my father for my eccentric ideas – he is, after all, a morris dancer. My colleagues think I have simply lost the plot; they take a wide berth around my desk and no longer engage in conversation, knowing that it will end up with me talking about bees.

It is no surprise, therefore, that I should reflect on precisely what it is that I am about to undertake. Especially when, a) I have spent my whole life running away from what I have always felt to be frightening insects, and, b) I don't particularly like honey. And yet regardless of these two small issues, I have started to learn the simple – or so I thought – art of beekeeping.

My decision to become a beekeeper started in the middle of the year on one of those fantastic summer evenings when the light is beautiful, resting on the garden, and I was there, glass of wine in hand, watering the flower beds. It was one of those moments to treasure until I realised I had completely drenched a poor bumblebee trying to seek shelter in the flower of a gladioli. The poor little thing did not look too happy but just bumbled along onto the next flower. I was transfixed, and sometimes it takes just a moment for me to become obsessed. This was a glorious creature just going about its duty when a great beast of a thing (me!) came along to interrupt its vital role in the great world we live in.

That moment got me thinking about the whole bee world and it was then that I started reading about the plight of the honeybee. I hadn't even considered that there was more than one type of bee (I now know there are over 200 different types of bee in the UK alone). It sounded as if they were having a hard time – and I mean a seriously hard time – and not just from the likes of ambitious and competitive gardeners watering their plants. Honeybee populations are dropping in considerable numbers due to a multitude of factors which have collectively been termed 'colony collapse disorder' and not a lot was being done, it appeared.

There were also other reasons why bees were starting to appeal. I was becoming increasingly fascinated by elements of the self-sufficient lifestyle and I love growing vegetables on the allotment. The old romantic in me had idealistic notions of taking my little boy up to the allotment, and each Saturday going to check the bees with him just to teach him about the world and where everything that ends up on his plate comes from.

In order to turn my idealistic thoughts into reality I had to start to learn the art of the beekeeper, if only to help the bees in my area. Maybe I could make a difference and cause a butterfly effect in the UK which would spread throughout the world and save the humble bee…

I made it my mission to learn everything I could about bees. I would get a couple of hives, bore my friends and family (even my morris-dancing father) with my new-found wisdom of the bee world and have a simple aim. Despite established hives being able to produce upwards of forty jars of honey per year, I only wanted to produce one pot of honey this year. Yes, that's right, just one jar of honey. It might not sound an awful lot but I have heard it can be rare for first-time beekeepers starting from scratch to get any honey in their first year. I hope you enjoy the journey.

SEPTEMBER 23, 2009

My beekeeping career started today with the first of ten two-hour classes. I found the beekeeping course by performing a Google search and discovering that there were beekeeping associations that ran evening classes. I was already starting to feel old even thinking about beekeeping, let alone thinking about attending evening classes.

I was feeling quite nervous as I drove to the local school where the course was being held, as I simply did not know what to expect. I was pleased to be earning brownie points as well as learning a new skill because, should we ever win the lottery, Jo and I would love to send our son to this rather grand school set in the heart of the Surrey countryside. Therefore, I reasoned, this was to be a reconnaissance mission as well as an evening class.

While driving along on this miserably dark autumnal evening, I was wondering how beekeeping could possibly take ten weeks to learn. Surely these little black and yellow insects would be easy to look after. I was more interested in what the fellow enthusiasts were like, let alone the teacher. I had a very clear vision, probably gleaned from my knowledge of morris men: usually old, with beards, red cheeks and noses, well-rounded tummies and generally a fondness for drinking ale. I felt that beekeepers and morris men would be cut from the same cloth. I wondered if being beekeepers-in-the-making, beginner beekeepers would only have partial beards, slight tummies and merely a hint of reddening of the cheeks and nose. The teacher, on the other hand, being fully qualified, would have all the attributes of the morris man.

As I drove into the school's vast driveway I was immediately in awe of the beautiful building in front of me, softly lit by floodlights. It was Gothic in appearance with impressive stonework and the most imposing arched windows and doorways dotted around its facade. I could just imagine Sebastian coming here. I approached the door of the classroom (which was one of the outbuildings and not so impressive, having probably been built in the 1960s!) with my heart beating slightly faster than usual. The strange nervousness of a new situation was dawning on me – as well as the frightening thought of a room full of morris-dancing beekeepers.

I opened the door and walked into the classroom. In fact, everyone looked pretty normal. Only about 40 per cent had beards – none of the ladies did – and there were only a few rounded tummies. They all said hello to me, which was nice. The classroom had desks laid out in two horseshoes, with a desk at the front. Having only just got there on time I was the only one sitting in the smaller, inner horseshoe with everyone else behind me. I felt like a naughty schoolboy having to sit closest to the teacher and voiced this point to the others to subtle smiles.

So the most difficult bit was done. Nerves gone, I just had to sit down and enjoy the next two hours. David, the teacher, was incredibly informative and immediately likeable. I hadn't spotted him straight away as he was standing off to one side. He was also the slimmest of the lot and had no reddening of the cheeks either, putting him way off my stereotypical beekeeper, though he did have the tell-tale beard. I later found out that he was one of the top beekeepers in our area. How do they measure this? Honey production? The beekeeper with the most beehives? Who knows, but I was certainly fortunate to be learning from him.

This first session covered the basics and gave an insight into the world that I was about to enter. Within ten minutes I realised why these courses were ten weeks long. There was so much to learn. I drove home from the session utterly in awe of what I had just learned. I now know what honeybees look like (they are not the fat, hairy bees which are so obviously bumblebees but in fact look similar to wasps but with not so harsh colouring) and realise just how important they are to the world in which we live. I got home, offloaded a load of (what I believed to be) useful information to my wife, and then remembered about the reconnaissance mission. I told her about the school: brownie points duly earned.

I can't sleep but I know I'm hooked on becoming a beekeeper.

SEPTEMBER 25

It is now two days on from the first day of the course that changed so many of my ideas about the honeybee and I find that I cannot stop thinking about them. One fact on my course amazed me and I feel I have to look into it a little more. Doing this will introduce me to the practical side immediately and make it all feel a bit more real.

I learned during Wednesday's session that bees can forage up to 3 miles away from the hive. This fact astounds me. Imagine the journey these little bees do, just in the search for nectar and pollen!

I am truly desperate to look at a local map but I don't want to rush into this. I have a notion of sitting down with a nicely brewed cup of coffee with a map spread out in front of me. I will locate where my hives are to be based (have not got a clue where yet) and get a pair of compasses and plot a nice circle around my hives to the tune of 3 miles. There is a side of me imagining a World War bunker-type operation, complete with the map sprawled out over the table, low-level lighting, cigarette smoke hovering overhead and me manoeuvring little bee models around the map with funny-shaped sticks.

I know I have a 1:25,000 map of the local area somewhere so I reckon this will be enough to tell me all I need to know. How many farms are there around here? How many fields for foraging and what types of crops are grown? This is obviously of utmost importance for the bees – I've heard that oilseed rape, for example, produces a very early honey harvest; if you leave it too long it goes rock hard apparently. I hope I don't have too much of that nearby. I feel fortunate to live in the country with lots of room for them to forage. I wonder if there's a difference between urban and rural bees and their respective honey...

SEPTEMBER 27

After a short trip away with work, which meant being out of the house at 4 a.m. and only just arriving home at 11 p.m., I have had enough of my corporate world for today and am very tired. I have worked in the overseas property business for four years now and it involves a lot of international travel. For the first couple of years it was fantastic but now that I can even tell when Gatwick Airport WH Smith has restocked its shelves the travelling has lost its appeal. However, with map in hand, I feel that now is the time to see where my little ones might fly to.

Jo and I get into bed and I bring with me the map and a glass of wine; who says romance is dead?! I also bring a pair of compasses ready to draw a nice circle around a proposed hive location to see just how far my bees will fly. I think this an ingenious plan though perhaps not the best implement to take into the marital bed. I then notice to my utter dismay that we are located right at the bottom of the map; I can therefore only see the top of the 3-mile circle.

Even though I can only see half the story I still know this is a huge area for my bees to forage – a total area of nearly 19,000 acres after some quick mathematics. I immediately realise why they say that bees literally work themselves to death. As I view the area I also realise just how little I know about my local landscape and, due to the fact that I live in pretty much the middle of nowhere, how little I know about the farming and agriculture around me.

I feel that I have to know more about this 'bee fly zone', and that I need to have a drive around to familiarise myself, not least because in that compass half-circle I count about five public houses. Imagine what I might find in the other, more populated half! As I am drifting off to sleep I feel it is entirely justified as maybe, just maybe, my bees might fly into the gardens of the public houses at some point and I might need to go and see what they are like. What a lovely excuse to go and investigate. A job for the weekend I think.

SEPTEMBER 30

It's the second session of the course tonight and again I come away with a great appreciation for the 'humble' honeybee. For such little insects they are unbelievably sophisticated. Essentially the topic for this evening is the colony itself and its structure, but I can see that David is itching to tell us all some amazing facts:

- In just one hive there can be up to 60,000 bees but just the one queen (!).
- To make one jar of honey (you know, your regular 454 gram jar from the supermarket) the bees from a beehive would have made at least 25,000 flights to gather enough nectar to convert into honey.
- The average worker bee, in their lifetime of only six weeks, despite flying for hundreds upon hundreds of miles, will only make one twelfth of a teaspoon of honey.
- At all times of the year, regardless of the outside temperature, the hive is kept at a temperature of between 32 and 35 degrees Celsius. It doesn't matter whether you are in the Arctic Circle or in the Sahara Desert!!

For someone who has been around bees for most of his life, it's inspiring to see that David's passion for them remains strong. So are his concerns. Though I understand that we are going to discuss bee diseases at a later date, he obviously can't avoid the elephant in the room: the problems bees are facing. I have read a few articles about the problems but I genuinely didn't realise their extent.

Currently bee colonies are being wiped out at a rate of at least 30 per cent per year, David says, every year. In some cases, beekeepers in the US have been seeing losses reaching 70 per cent in some years. The almond plantations in California are already having to ship in beehives to help pollination as there simply aren't enough bees to do the job locally. Considering this is an 800 million dollar business there is a serious dependence on bees: can you imagine manually pollinating thousands of acres of almond trees? I have heard about a situation in the deepest depths of China where people are employed to walk around orchards all day with feathers on long sticks to manually pollinate fruit trees. I can't quite see this happening in America somehow. Meanwhile, shipping thousands upon thousands of hives could be contributing to the problem, with the bees getting stressed on long journeys.

What is also interesting is the breakdown of the colony. Of the 60,000 bees in the colony, 90–99 per cent of those are the daughters and these are termed the worker bees. The name is particularly relevant when you consider what these bees do in their lifetime:

- Clean the hive and other bees
- Feed the larvae, young bees and the queen
- Deposit the pollen and nectar brought in by older, flying bees into cells and start the conversion to honey
- Maintain the hive's temperature by either huddling together in winter or fanning the hive in summer

 Make wax to build the comb

 Guard the hive from intruders

Incidentally, this is all before they are old enough to leave the hive, about three weeks after they hatch. They then simply work to bring in nectar and pollen for the hive, before dying of exhaustion out on the wing; therefore reducing the work of the others back at the hive. David mentioned that they are the perfect example of a successful democratic society and I can see this already. They all work together for the good of the hive: incredible, really.

It sounds a pretty tough life, especially in comparison to their brothers who seem to have an amazing life! The boy bees are called drones and when I saw a photo of one against a worker, it was like watching an episode of *Supersize vs Superskinny*. The drone is almost twice the size and is essentially a fat, lazy slob of a bee. The drones simply wander around the hive expecting to be fed, cleaned and generally treated like royalty. Their sole job in life, other than just chilling out, is to mate with a queen. Somehow they know when a queen has left a hive (how do they know that?!) and the drones fly off to a secret location and compete to get their wicked way. Apparently the queen may mate with up to seventeen drones – she must be exhausted after that! If the drones succeed and are one of the lucky ones able to mate with the queen they do meet a pretty swift end. While mating, there seems to be a point where their enthusiasm gets the better of them, as their abdomen splits in two and they die. If they don't succeed in mating, though, they are still alive – I should think they fly with their proverbial tail between their legs back to the hives.

If they don't manage to mate with a queen by the end of the summer season, says David, their sisters, the workers, get the hump. In short they get their wings nibbled off and are booted out of the hive. As they cannot fly off anywhere without wings, they have a miserable end as they succumb to the elements. Therefore, it must be said, they have the most amazing lives but also a rather quick and untimely end!

OCTOBER 3

I find myself at work today daydreaming about bees, which feels a little weird. I am contemplating my understanding of this new world, how little I knew before and how amazing it all is. In just two sessions I feel my taste buds for a new hobby are burning. Never did I think I would want to be known as one of those slightly strange beekeepers, but I can feel I am turning – I know what I'm like. I am most likely to become obsessed. What will my friends, family and colleagues think? I think I will wait some time before telling them my plans for the year.

This concern all stems from a rather tenuous link from my childhood, I think.

I used to have various money-making schemes to raise cash to spend on comics and my addiction to penny sweets; cola bottles and fried eggs in particular. To complement my pocket money I would wash cars and do odd jobs and gardening for people in the local area. I remember once putting little leaflets advertising my services in people's letterboxes to help finance my addiction to *The Beano* and *The Dandy* while scoffing flying saucers.

One of the people that responded was Anne Buckingham, who my parents always referred to as 'the lady who keeps bees and chickens at the end of the road'. Her car was a grey Saab with the most amazing windscreen – almost vertical but fabulously curved. Washing her convertible grey Saab was one thing, but I distinctly remember peering through soap-sudded windows and seeing her looking rather funny in an all-in-one white boiler suit at the bottom of her garden.

I will always remember laughing as this lovely lady with rather unkempt hair pulled on her boiler suit and week after week fell over trying to put on her wellington boots. She would then trudge along to her two beehives, tripping over her own feet as she went. When she reached the hive, however, it was a different story. She became calmness and patience personified as she went about her business, with a metallic object puffing smoke at the bees. Still, 'utter madness,' I would think as I went about my weekly task of removing droplets of pollen from her car chassis.

Beekeeping to me as a child was therefore carried out by middle-aged, Saab-driving ladies with an amazing ability to fall over their own feet. This viewpoint never really changed into my adult life, and thus the hobby never really appealed.

Until now... Heaven help me!

OCTOBER 7

I am sitting here in my study after a long day in my corporate world, exhausted as I had to do some travelling last week and haven't really caught up yet, followed by my third beekeeping session. Tonight's session was about the beehives themselves – and here was me thinking there was just one type. For the first time

I have started to imagine my own beekeeping next year, and to consider what hive I will get. I really have to think this through to make the right decision.

Previously, I thought beekeeping was simple. You would put this beautiful white beehive, looking a little bit like a pretty version of a dalek from *Doctor Who*, in the corner of your garden. When you were ready you would pop over and use the tap on the side to pour some honey in the jar, before walking jovially back to the breakfast table to spread it on your toast.

In fact that good-looking dalek, which tonight I found out was called the WBC hive, is rarely used now. William Broughton Carr designed it (hence the name) in the late 1800s and it quickly became the quintessential British beehive. However, it was forty years earlier that the first what they now call 'moveable frame' beehive was patented by a Rev. Langstroth over in America. It's apparently the world's most popular beehive today, with over 75 per cent of the world's beehives being a Langstroth. I hope he signed a royalty deal.

Reading about this session before the course started, I did wonder exactly how interesting this evening about the hives would be. But I have to say I have been pleasantly surprised. I never realised just how long beehives have been used, and it is quite amazing to think that beekeepers still use a piece of equipment that was patented over 150 years ago, with no major changes. We can't say that for many things nowadays, can we?

We also talked about a hive called the 'National'. Being British, I suppose we wanted a bit of our own engineering and essentially we have ignored this popular American Langstroth design. The National, a smaller version of the Langstroth, is the most-used hive in the UK and so maybe I should look into using one of these.

I am not convinced though because I never follow the crowd, and I am therefore not 100 per cent certain that using a National or Langstroth is right for me.

David also talked about more modern hives; some being polystyrene and some being made of plastic. It all sounded a little strange to me and the feeling accelerated when I saw pictures of them. The plastic hives, called Beehaus, looked a little bit like top-loading freezers but were all bright colours, yellow and purple. They did catch my attention.

David did not sound the biggest fan; he stated that most beekeepers dislike them. I need to know more though, especially as they are compatible with the National hive that had been previously recommended – one of the most important factors if you are considering two or more hive types. Somehow David's hesitation to recommend it fuelled my interest, as I always like to give everything a fair trial.

As a beginner, said David, you should look for a hive that is compatible with other local beekeepers so that in times of emergency they can help you out (I hadn't a clue what that meant if I am honest) and that, most importantly, you should also run two hives so that you can assess each colony individually and have a comparison.

Oh Christ, not just one then!

I now have to convince Jo that I will be looking for two colonies of bees, which could mean up to 100,000 bees; and am no longer looking for this beautiful WBC hive but two completely different hives, one of which looks like a brightly-coloured freezer box. Hmmm... This could be interesting.

OCTOBER 10

After doing some research these last few days, I have decided on my hives. I am going to compare and contrast two hives. One will be the traditional hive known as the National; however, I have decided to go for the larger version, more regularly known as the 14 x 12 which refers to the dimensions of its frames: 14 inches wide by 12 inches tall. Essentially, this is just a little bit bigger than the usual National. Apparently, due to selective breeding, we have prolific egg-laying queens in the UK and this combined with the warmer summers is resulting in larger colonies, so it is recommended to use these larger hives. Overcrowding is one of the commonest reasons for bees swarming early on in the season – and I really don't want that to happen if I haven't got a clue as to what I am doing.

It can be a little bit more difficult to handle at the height of the summer what with the larger frames, they say. A thin wax 'foundation' strip about a millimetre long is connected to a wooden frame and inserted into the beehive. This forms the basis upon which the bees build their comb, in which they put pollen and nectar, and the queen lays her eggs. The resultant weight of its contents and the bees themselves can break the comb when you are lifting it out of the hive, which can result in being covered with bees. If you have seen the Eddie Izzard sketch aptly named 'I'm covered in bees', this is what I am assuming will happen.

I will compare the National hive with the oversized, brightly coloured freezer box. Everyone seems to hate it and so I have to give it a go and see if it really is as bad as everyone makes out. Apparently it is based on an old design called the Dartington hive, but is made out of plastic rather than wood. One thing that is attractive about it is that it is actually two hives rather than one, side by side... But then there is a temptation to have three colonies, rather than two... Help! This is getting addictive!

The other thing about this Beehaus is its marketing. I have to say I have fallen for its tag line: 'With a Beehaus in your garden, you'll soon be saying "Show me the honey!"' How can I deny a space in my garden for this hive if it promises to produce that single jar of honey I am looking for? I can only imagine the bemusement of my neighbours next year when I am shouting to the bees 'Show me the honey!' at the top of my voice like Tom Cruise in *Jerry Maguire*. As if beekeeping wasn't bad enough, imagine a beekeeper who tries to entice his bees into producing more honey by quoting famous movies at them. Do gardeners do this to produce prize courgettes?

OCTOBER 14

I am sitting here with a nice glass of red wine reflecting on how I never quite realised the long history of beekeeping. At the last session, it was evident that we still use equipment that was introduced back in the 1850s. But man was dependent on the honeybee well before that.

Back in Egyptian times, the Pharaoh himself was the god of honey and honeybees were seen as teardrops from the sun. Honey was also used as currency by the people of ancient Egypt in payment for land rents, and detailed reports were kept of production and payment: the first evidence of organised apiaries.

With reference to the UK in particular there is documented evidence dating back to Roman times and then Anglo-Saxon and Norman times of widespread beekeeping. In a rather cold schoolroom we are learning a hobby for fun that for a long time was very, very serious business with large financial, religious and social considerations. I feel a little bit humbled and think that I should be taking this a lot more seriously than I have started out doing.

It is also quite clear that honeybees have been around far longer than us. And yet now, after many years of exploitation and manipulation to extract as many resources from the hive as possible, the bees are suffering. It makes me feel a little sad to be honest.

OCTOBER 17

While browsing around the Internet for bee-related things, I came across architectural plans for all sorts of beehives and it has got me thinking. Hives aren't exactly cheap and so maybe I could just make myself a hive rather than buy one. It doesn't look too hard to do; after all, essentially it is just a wooden box. The difficult bit looks like it will be the joints – quite fundamental, you might say – and then what they call the open-mesh floor, the bit at the bottom of the hive that the box sits on. It is an open mesh to allow ventilation through the hive but also has some beneficial disease prevention reasons behind it.

I can be quite sentimental at times and so am thinking about trying to get my father involved; that way all three generations of Dearsleys could be involved in my bee exploits. I have fond memories of helping Dad when I was younger. There he was in his workshop, otherwise known as 'the cold bit at the back of the garage', working bits of wood using an elaborate collection of hand tools – never the newfangled electrical gadgets. He would always have his pipe hanging loosely from a corner of his mouth, smoke just dribbling over the sides. Every so often he would stop, stand upright and, while looking up through the only window, remove his pipe, cupped in the palm of his hand, and exhale a dense cloud of smoke. I loved those times and I thought Dad was the world's leading woodwork expert.

Well, despite his knowledge of working with wood, if you look up the word 'bodge' in the dictionary, my father's name is there enshrined in history, and so it may not work according to the plans.

I seem to have inherited this 'bodge' gene, if there is such a thing, but I am working on the principle that two negatives make a positive. Therefore our two bodge characteristics might work together well and we will produce a fabulous-looking hive.

Note to self: broach this idea with Dad. It would be great fun to do this together.

OCTOBER 19

The hive-building day is on!

Having loosely discussed the idea, Dad is willing to help out. I wouldn't say he was jumping-over-the-moon keen, but I suppose it isn't every day your son rings to suggest building a beehive together. I have a feeling he is still in shock that his son is becoming a beekeeper. His dreams for many years of me becoming the next champion morris dancer must be slowly ebbing away as I don a different kind of uniform, with no bells in sight.

We have set a rather random date of 6 March to have it all built as I figure my bees may arrive around that time, since that is when the season supposedly gets going, or so I have been led to believe. It should also give both my father and me time to order some of these plans which are readily available online and then order the appropriate wood to have a few trials. God knows what wood I will use, as again it seems there are many different options and nothing is straightforward. No doubt most of the trials will be complete bodges so I just have to get to a competent level of bodge before Dad and I attempt a final sample over that weekend. It's all very exciting.

OCTOBER 21

It has been dawning on me for a while that beekeeping is a little bit more involved than I first thought. Tonight's session reinforced this as we discussed the beekeeping year. Who would have thought it all revolves around a cycle, the same each and every year?

It was fascinating to learn that at the peak of the summer there could be 60,000 bees in a hive and yet a few months later these numbers will have reduced to around 5,000. By my own amazing mathematic ability it means that there are 900 more bees dying than being created every day for nearly two months.

Gradually, as winter turns to spring, the queen will begin to lay eggs and the colony gets going again. Soon she starts to lay towards 2,500 eggs a day; her maximum capacity and more than her own bodyweight in eggs every day. Obviously as a result, the colony expands rapidly.

As a beekeeper you need to tend the colony once a week, usually about half an hour per hive from May through to September in what is called the peak season; you may check more sporadically in the months of April and October, but only on warm days, and during the winter months you are allowed to enjoy yourself and put your feet up while having honey on toast, under beeswax candlelight while enjoying a glass of mead. Must find out more about mead as it sounds delicious!

David also mentioned that if beekeepers are really lucky there could be two honey extractions per year. So it's not a constant stream of honey, as I had thought. There could be one in the spring if you have a strong colony coming out of winter and you have a good amount of early flowers or fruit trees nearby. The usual and more expected harvest is in August after the main 'honey flow'. He also said that it was very rare for a first-year beekeeper to have a good crop of honey as the colony may not be strong enough.

Hmmm... I wonder if I will get any. Just one jar, please!

OCTOBER 22

I ordered some hive plans over the Internet today – all easy to do and very cheap. Something tells me though, having viewed the document online, that it might not be as straightforward as I had previously thought.

OCTOBER 25

My hive plans have arrived. I love the way that the first line says 'competent woodworker required'. I have to say, looking at them, they are not particularly easy. Essentially, all the plans do is provide very exact dimensions, rather than actually telling you how to put the parts together. That is like giving a cook all the ingredients, and then letting them guess how to cook it all. I am not sure how successful I will be at this.

Having put the plans on the sofa I then watched as Sebastian crawled over and pulled himself up to grab the plans before plonking himself back down on the floor with a thump. In a way that only babies can, he then proceeded to read the plans upside down while trying to eat one side and tearing the other. The bemused look on his face as he was attempting this major feat of childhood mirrored my feelings for the plans themselves. His face was a picture and I knew exactly what he was thinking.

OCTOBER 28

At tonight's session, David is running late and we all get talking for the first time. The initial awkwardness of not knowing each other has gone and we have a common interest which makes things easier. There's a real mixture of people getting involved. I find out more about my co-learners. There's the father and son who have been beekeeping before but want to have an update on modern techniques (when I say 'father and son', the son is easily in his fifties). There are a few others who have kept bees previously, including an Aussie guy who sounds quite experienced in keeping bees but only in Australia, who wants to find out why we are better beekeepers (OK, maybe I'm making that bit up). Then there are a few, like me, who just want more information but are keen to get started. There are at least three married couples; it is evident that one of the pairing is keen and the other, duty-bound, has come along for support.

The average age is about 45 and about 60 per cent are male. Thirty per cent of them started the course with beards and now I would say at least 50 per cent have them and so it does seem that some are starting to morph into my stereotypical beekeeper as the course goes on. Fortunately I don't seem to be changing just yet, although I do seem to be enjoying more cider recently...

OCTOBER 29

Last night at my session not only did I learn a little bit more about my fellow classmates, we also learned a lot about swarming. Before now I had never considered swarms, other than hearing horror stories about people seeing them fly by with almost military precision, with a noise equivalent to a jet plane flying past, and everyone diving for cover. Needless to say I was pretty sure swarms were not a good thing. I was quite taken aback when I realised it was quite the opposite in fact.

I had never before considered the reason for a swarm. I discovered that it is an example of the amazing perception of the colony that knows it is under threat and does something about it. A 'new' queen is raised – how, I do not know yet – and the old queen leaves the hive with between 1,500 and 30,000 bees to set up a new colony elsewhere. There are a multitude of reasons for this, which could include a diseased hive or the fact that they are running out of space but in any case, they do it in the interests of the colony.

It also transpires that this is the time that people are least likely to get stung. David showed us pictures of beekeepers with various limbs being inserted into a swarm once it had come to rest. He then decided to show us a picture of a beekeeper with a 'bee beard', which is exactly what it sounds like, i.e. thousands of bees that affix themselves to someone's face. As they are in a swarm state they are said to be calm and docile. Having looked online tonight, while thinking about this whole swarming malarkey, I found out that the current 'world record' – how is there a world record for this sort of thing?! – was 57 pounds of bees!

The crazy thing about these bee beards is that they date back to the 1700s. Surely there must have been better things to do than layer your chin with bees. Another ancient technique was known as 'tanging' – apparently back in the day, people saw a swarm as good luck and hence tried to lay claim to the swarming bees. They would run after the swarm banging pots and pans to try to calm the bees and 'tempt them into stopping'. People would then rest their handkerchief over the swarm to lay claim; another piece of silly British tradition and a practical demonstration of just what you can do with a handkerchief.

NOVEMBER 4

Tonight it really struck me just how much I have already learned about bees but, at the same time, just how much more there is still to learn. David took us through the mechanics of the queen bee today. It just shows how important she is with a whole session dedicated to her, and to be honest she is quite an amazing subject. However, David was quite quick to state she was simply an 'egg-laying machine', and though it was a relatively complex job, that was all she really was.

I can't quite imagine our queen wanting to be labelled this way but it was interesting to hear that actually the queen isn't the real leader in the hive. Like us, a democratic society, the workers and drones are the real decision-makers (OK, I realise that might sound a little naive!) and arguably, because of the variety of jobs they do, are also the more advanced bee.

The queen bee seems to keep the colony together and calm by emitting pheromones. Apparently, if the queen suddenly dies, within fifteen minutes the colony will be aware of this and will immediately set about raising a new queen. This is also true if they feel that the queen is losing a bit of strength or if she has accidentally been damaged; the bees will start raising a new one, even if she is still present in the hive. I find this all rather astounding. How on earth can a colony of 60,000 make a collective decision on these sorts of matters within fifteen minutes?

I have a theory that it is either a very complex game of Chinese whispers (although unfortunately if that were true, what started out as 'we must raise a new queen' could end up something completely different) or evidence of a highly functional, structured and organised set-up which, at my current level of expertise, I simply cannot explain.

Can you imagine this happening in our world: 60,000 people trying to make a decision to essentially bump off the Queen? It would take fifteen years, not fifteen minutes. I couldn't see Queenie being too pleased if, while walking around Buckingham Palace, she saw one of her footmen desperately hiding a new queen behind a coat of armour in the corner of the throne room. In the bee world, the old queen gets the hump and flies off with half the colony. To top this off, David went on to tell us some even more amazing facts about queens that I wasn't already aware of:

- A worker will only live for about six weeks whereas a queen can live for up to five years.
- After her mating trip, the queen will keep laying eggs for the rest of her life at a rate of up to 2,500 eggs per day.
- The queen can select whether she fertilises an egg or not – if she fertilises the egg she creates a worker, if she chooses not to, a drone is the result.

All in all it was a pretty fascinating evening; so much so that as I left the classroom in deep thought, I managed to fall down all the stone steps to my car. This happened just as another group of people were walking out of the main building only to see me perform a stuntman-like somersault down the steps and land on my feet. It must have looked amazing aside from the fact that I landed on wet leaves and so skidded along before promptly falling on my derrière. Not my proudest of moments! But I did say I was clumsy.

NOVEMBER 7

I must be addicted. Never before in my life have I ever taken homework seriously, but on Wednesday we were given the task of reading some leaflets about bee diseases ahead of next week's lesson, and here I am tonight dutifully sitting in front of the fire with a lovely glass of red wine (maybe that is the difference from my school days) reading the leaflets word for word. My God, bees are not having a good time of it; my God, there are so many diseases.

Shockingly, not only did I do my homework but I also found myself reading around the topic; something my parents and teachers could only have dreamed about when I was a child. I was going online to find out more about the diseases just so I was better prepared for next week... Quite scary really, but I am already excited about starting next year.

NOVEMBER 11

I knew tonight was going to be a rather sombre occasion as I had learned about the diseases but I never realised quite the impact it would have. I would advise you now to go and get a nice strong drink to prepare yourself for a rather melancholy read!

I was expecting to hear that bees were getting the equivalent of the human cold and that reports in the media were being slightly exaggerated; such is my optimistic attitude to life. What I wasn't expecting was the fact that for once our media are rather downplaying the problems. It is more like Armageddon for the global bee population as a pneumonia virus sweeps through it.

David was very good at explaining the issues but the frustrating thing for him as a bee inspector and, from the sounds of it, for every beekeeper alive, is that there is no complete diagnosis. It did get a bit technical so my revision certainly paid off. In brief, it sounds as if the bees' immune systems are weakened as larvae, probably by a mite called varroa. This is a vicious little bed-bug-like mite that, if seen up close under a microscope, would give children nightmares for weeks. They weaken the larvae such that, as adult bees, a whole host of secondary diseases make their move and kill them off.

From the sounds of it upwards of 30–50 per cent of hives are being affected every year at the moment, with colonies literally 'collapsing'. Apparently beekeepers are finding hives either abandoned, with no bees in them at all, or there is a slow and gradual decline in numbers until they all die a painful death, unable to look after themselves.

It seems pretty desperate. I needed a drink after the session and so popped to the local pub with some of my new beekeeping buddies. Some of them who had kept bees before had experienced colony losses themselves. It is amazing how attached people get to their bees and obvious how upsetting it could be to see them simply disappear.

A pint seemed to nullify the feelings of sadness at the situation and made me more determined to do what I can to help. Back home now, though, the enormity of the situation hits me again. I feel a call to arms is needed! Hence I have decided to set up a Facebook page for other beginner beekeepers (www.facebook.com/beginnerbeekeepers) to see what or who is out there. It will be nice to be able to speak to other beginners out there, to share experiences however good, bad or – in my case – stupid they may be.

NOVEMBER 15

A few weeks ago, you may remember, I sat in bed contemplating using a pair of compasses to measure my bees' flight path on a map. This was before I thought better of it as I was lying there next to my sleeping wife. I would definitely have been in the dog house if I dropped the compasses and stabbed her in the back while she slept. Quite pleased I didn't go through with that plan in hindsight.

Even using those rough estimates from before I still can't quite believe the size of the area in which they fly. I also can't quite believe that despite the fact I live in the arse-end of nowhere, with cows and sheep for neighbours, there are sixteen pubs within the area. Today I was able to put the two maps needed together and review the pub situation more closely. What an excuse to have a drive around, especially as I have never even seen half of them, let alone been in them.

Suffice to say Jo and I took Sebastian, our very little but ever-so-chubby one, on a ~~pub crawl~~ road trip. Essentially we went out on the proviso of visiting a couple of pubs in the local area for a few drinks and a spot of lunch. It has been quite a while since Jo and I have ventured out because Sebastian is still young, so this was deemed a real treat. Well, it was until I requested that we drive the 'long way round' so that I could look out of the window at the fields. I feel it's important to understand what my bees will be foraging on locally. It had never really dawned on me before that different flowers or crops would produce different honey and also require different methods of managing a hive. Without knowing what was growing locally, it would make the job that little bit harder.

Imagine realising that your husband wants to visit the pub as a cover for driving around looking at fields for an hour or two. There I was, notebook in hand, nose literally stuck to the window as Jo drove around, Sebastian asleep in the back, writing notes on all the fields I saw. How very sad. A trainspotter is one thing but a field-spotter is quite another.

The worst bit was, and I should have realised this before we set out, we are in the depths of winter. What hope had I got of knowing what was planted? It was immediately obvious that yes, there were lots of fields, but most of them contained 6-inch-high stubs of previously harvested crops. It was either that or freshly dug-over soil for mile upon mile.

So I learned a lot today but more about the local landscape, and a few pubs, than I did about what my bees might be flying to. I will say that the afternoon got a lot more fun after the third pub, having my third variation of local ale – especially as it was a Christmas beer called 'Santa's Wobble'. As the name suggests I was wobbling slightly as I left.

Still, it's certainly a job worth doing, though the mission wasn't really accomplished; maybe I will just have to do it all again in the spring. I might have to work harder at convincing Jo next time, though.

Strange as it seems, in a couple of days it's the penultimate session of my training course. I still haven't seen any bees and yet I feel I am becoming strangely attached to these little black and yellow insects that for years I have been afraid of and tried to run away from. The fear seems to be abating the more I understand them and the important role they play in all of our lives but I realise it will still be a minimum of four months until I actually get to see my own bees and get my own hive. It seems a long way off.

NOVEMBER 17

Today's penultimate session dealt with the 'products of the hive'. Here I was thinking that meant basically 'honey'.

I had already learned that worker bees have little wax glands on their back. The generated shards of wax are then moulded and manipulated to build the wax cells to deposit the honey in or for the queen to lay her eggs in. I know this sounds funny but I hadn't put two and two together and realised this wax can then be melted down and made into beeswax candles – I'd never really linked up the name before now.

There are also some beekeepers who specifically harvest pollen and attach so-called pollen traps to the outside of their hives – rather ugly-looking, brightly coloured boxes which knock off pollen from the backs of the bees' legs as they fly into the hive. I have images of little boxing gloves attached to springs which come out and punch the legs of bees as they walk through the trap. Apparently some beekeepers sell local pollen for people to eat. 'A teaspoon a day keeps the hay fever away', to take a popular phrase and change it around a little bit.

Then you have royal jelly, which really sounds special. Apparently royal jelly is fed to eggs and larvae to provide a rich diet of pollen and nectar; eggs selected as workers are fed it for a few days before switching to another foodstuff, whereas eggs selected to generate a potential new queen are fed royal jelly exclusively. Now I had heard about royal jelly before in hair shampoo but didn't have a clue that it is essentially bee food. I certainly didn't realise that it had great medicinal qualities. It is used to control Graves' disease and stimulate stem cell growth, not to mention its cholesterol-lowering and antibiotic properties. I believe, having heard all of this, humans should bathe in royal jelly every day for an hour or at least use it as an alternative to ketchup, and we would all be much healthier. I know there are some beauty salons that already use bee venom in some procedures to make people look more beautiful so I wonder how long it will be till health farms start to offer these royal jelly baths. What a great present for your wife – a bath filled with bee food!

There are plenty of other products that can be harvested from the hive – propolis, or bee glue, is another one that beekeepers can sell on for ridiculous sums of money in some countries – who knows why? Perhaps it's simply due to its scarcity and difficulty of extraction. Until today I had just considered honey as the sole product of a beehive and was simply thinking of a beehive as something that looked nice at the bottom of the garden. I have never before considered the huge variety of substances that bees produce and the ways and means of extracting them.

It was quite interesting to see the opinions bandied around in the meeting, though. Some of us had fallen for the idealistic notion of beekeeping and helping the bees out of this spot of bother they had got into recently. Others were looking at it through commercial eyes as a money-making option in these hard economic times. I have to say, it does seem a little strange to exploit what bees are producing as surely there must be a reason they make it all in the first place. I cannot see that one or two jars of honey are too much to take away but I am sure taking all of the products away from the hive cannot be a good thing. They certainly don't make it for our benefit and for us to take away from them. It seems almost wrong that they are struggling and yet we are harvesting everything they produce for themselves. That cannot be right, surely?

Food for thought.

NOVEMBER 24

The time has come. Tonight was my last session, time to bid farewell to this group of people that I have come to know through a mutual interest over the last few weeks. Who knows if any will gravitate to becoming a true beekeeper and take on the practical element of the hobby next year but I know one thing's for certain – I will be.

David is a bee inspector and it's my understanding now that he is one of the revered few who know exactly what they are talking about. Should I join the British Beekeepers' Association (BBKA), he might pop up during a hive inspection to keep an eye on what I was doing. Daunting as it sounded, tonight David turned into a salesman and recruiter for the local beekeeping association.

For the end of the course, we had a film to watch. It felt a bit like the last day of school when you were a child and you were able to play games or watch the TV. What fun! Halfway through there was a knock on the door and in entered, in my mind, the most beautiful specimen of a beekeeper. The beekeeper I had always imagined. The beekeeper that maybe one day I will be. The beekeeper that everybody knows. The beekeeper that looked, to me, like a morris dancer! It immediately set me at ease. My original stereotypes might be real after all. Here in front of me was a real beekeeper.

Andrew walked in: a bit dishevelled, aged about sixty (I hope that is kind if you are reading this Andrew) with a full-on beard; it was a beard that any man would be proud of, sculpted yet disorganised, fluffy yet manly, the colour and consistency akin to that of Father Christmas. I think I had beard envy. Andrew had little rosy cheeks, though I have to say it was probably due to the cold weather outside rather than an abuse of local ales or cider. He also had a rather large belly, one I would expect of a beekeeper, and a lovely and jolly character.

I knew this was all an act and yet Andrew and David made the perfect double act, lulling you into joining a local association. It must be said however, it did sound like the right thing to do, especially as it meant there was a wealth of knowledge at your fingertips and people to share experiences with locally. No doubt I will need this next year.

I didn't need too much convincing. Here I was standing in front of my idea of a real beekeeper, and I was happy to do whatever he recommended. I pretty much filled out his rather crumpled up and damp forms while he was there but thought I'd better speak to Jo first. Andrew made his exit into the night, obviously satisfied that he had bagged a few more recruits, including a 'Young One' as he called me in obvious delight when I took away his information and membership form with such enthusiasm.

We watched the rest of the film summarising the course and it was actually the first time I had seen bees in action and beekeepers working with them. It was something to behold, thinking that in a few months, that could be me.

The paperwork David was filling out while we watched turned out to be our certificates. At the end we had a ceremony to certify that we had all attended the course. It felt a little like a passing-out parade as we all shook hands with the inspector himself and he wished us luck. We all said our polite goodbyes, left the classroom and headed into the cold dark November evening, certificate in hand, feeling just a little bit more the beekeeper than we were when we first walked into that classroom ten weeks previously. As I walked away looking at my scroll of paper, through the magnificent surroundings of this beautiful institution, alongside some other wannabe beekeepers, I only wished that school had been like this. However, our impending visit to the local pub definitely confined those wishes to the grave, especially as I would get served without question these days.

Still, I am now some way to becoming a beekeeper and I am desperate to get my hands dirty.

NOVEMBER 26

The nights are drawing in so quickly and the days are so short there isn't a lot of time to do anything. We put Sebastian to bed and I have come downstairs to think about where I go with my beekeeping. As I sit on the sofa next to the window staring out into the black, only to see my reflection looking back at me (I am not one for pulling the curtains closed too quickly at night), it seems strange to think I have been whipped up into a frenzy of excitement and amazement about these little insects, only to have to wait for a few months before I can actually do anything practical. There is no way of getting bees this late on in the season as it is just too cold outside. According to David, you generally obtain bees two ways: by receiving a swarm, or buying a small hive of bees from a recognised source. Either way, this won't happen until late spring at the earliest so I have to temper this excitement for now and do as much reading around the subject as possible.

Part of me wishes there had been a spring course so I could have gone out and got involved in the practical elements immediately. Heaven knows where I am going to get my bees from, regardless of whether I opt for a swarm or recognised supplier. Another great reason to sign up to the local association. I shall try to have a chat with some of the people there.

One thing I am definitely aspiring to next year is this one jar of honey. Apparently, if you get your bees early enough, it shouldn't be too hard, though it isn't a given. Knowing my luck I will probably make a pig's ear out of it and stand no chance of getting that jar of honey at all. If I am lucky enough to fill a jar next year, I should imagine I will celebrate by sharing a freshly toasted slice of bread dripping with my honey with Jo and Sebastian. I couldn't think of any better way to mark the end of the first year.

I find myself looking around in supermarkets at the honey and considering where it comes from. By this I mean geographically rather than the obvious origin of a beehive, although I would put a bet on the fact that some honey has no link to bees or beehives whatsoever given our ability to create artificial foodstuffs.

As I look out onto my now-dark garden I can only imagine how wonderful it will be to taste the honey from my own garden. It must be a lovely feeling, knowing that the honey you have on your toast is coming from your own flowers. I wonder if you can taste the flowers. It sounds a strange thought, but as I now know you get different sorts of honey, I wonder if there will be a particular 'James's garden honey' taste.

I am now resolute, my mind made up, whatever the cost I will make a jar of honey next year. I shall stand next to the hive and order them all to fly just that little bit more, to work just that little bit harder in the hope that I can save face with everyone and enjoy just one single jar! If I am doing this beekeeping malarkey, I have to consider the bees first, of course – but consider my breakfast table a very close second.

DECEMBER 2

Today I managed to sit down and read some of the information I was given by the Reigate Beekeepers' Association which all seems pretty comprehensive. I never realised before what the membership entailed but I am really quite impressed. The cost of membership, for all the added benefits, seems quite reasonable. I am entitled to attend all of the summer and winter meetings and lectures and by the looks of things there are many, not all of them terribly enticing, ranging from AGMs and EGMs to Australian beekeeping talks and candle-making workshops – now what would my friends think of me attending the latter? One did catch my eye though: making mead. That might be a more acceptable course for me to attend and even to gain some respect amongst peers; perhaps after they had enjoyed a few glasses of my homemade brew.

The fee also includes insurance for up to three beehives; something I didn't realise was required. Apparently you need it on two levels. Firstly, your bees might sting someone who could take issue with it and sue you – though how on earth could anyone prove it was your bee in the first place?! I suppose if they were standing next to the hive then yes, but I can't imagine someone being stung 3 miles away, seeing the dead bee on the floor, examining it and noticing that it had the hallmark of a Dearsley bee! It does seem crazy that you now need insurance against 'your' bees stinging people. What is the world coming to?

Secondly and on a more serious note, the insurance provides compensation in case you lose your hives. There are some bad diseases (American foul brood is one such disease) where you have to inform a government department immediately. They will send over an inspector who, if it is confirmed, will dig a hole in the ground, put the hives into it and set fire to them. What an awful sight that must be.

Insurance aside, I know that joining the association is the way to go. For my beekeeping experiences it is obviously right but, personally, I still know very few people locally despite Jo and me moving here over four years ago. This will be a nice way to get to know the locals.

DECEMBER 5

Just realised my membership form is still on the bureau by the door. Note to self: must put a stamp on it and post it!

DECEMBER 9

Jo must have finally got fed up with a letter clogging up the bureau and posted my membership form and cheque today! I wonder what happens from here. The last time I joined a club was when, as a child, I joined the Dennis the Menace Fan Club for about six pence.

I remember jumping for joy as a Dennis the Menace-themed envelope dropped through the letterbox. For an eight-year-old child it was the equivalent of Christmas, receiving a red and black envelope with your name on the front. Inside were two badges but one was extra special: Gnasher's badge which was furry and had those eyes that sat in a little clear Perspex lens and moved around when you shook the badge. I remember wearing it for weeks on end and never wanting to take it off, telling all my mates that I was now a member of the Dennis the Menace Fan Club.

Not so sure that the Reigate Beekeepers' Association will be sending me through a yellow and black striped envelope with bee-themed badges any time soon but there is the same level of anticipation my end. Having not joined a club for nearly twenty-five years this feels like a big moment for me!

DECEMBER 20

Jo and I have always loved Christmas and this year seems all the more special as it's our first as parents. The tree has been up for a while now, much to Sebastian's interest; he just lies there looking up at the sparkling lights, trying desperately to reach out and grab a bauble. I should imagine it is far better than looking up at the usual array of soft toys.

For some reason we have decided to cook the family dinner this year. The idea was simple: we wouldn't have to travel with Sebastian, which could have been a challenge. However, it does mean there is the rather complex thought of planning the Christmas dinner for around ten people, all of whom are excellent cooks in their own right. Since Jo will no doubt need to feed Sebastian mid-basting, it means cooking duties are left to me. Heaven knows what will happen and whether the Yorkshire puddings will rise or fall flat, whether the roast potatoes will be crunchy on the outside and fluffy on the inside and whether the gravy will be 'just right'. I am starting the planning now!

With the decorations out, we are getting ready for the arrival of Father Christmas for Sebastian, and I can only think that it is a magical time of year. It's all the little things about being a father at Christmas which I haven't experienced before that I am looking forward to. Perhaps the highlight will be helping Sebastian put a carrot and mince pie out for Father Christmas and Rudolph by the log stove. I am going to mention, obviously, that Father Christmas will be tremendously thirsty after such a long journey. I am sure Sebastian will help me to pour a small drop of port to quench his thirst and give him a little bit of warmth in that rather rotund tummy of his.

Thinking of men with beards and round tummies… I wonder what bees do for Christmas? I wonder if they all gather round the centre of the hive for a specially prepared Christmas dinner of honey and extra-special pollen titbits.

DECEMBER 25

Father Christmas dutifully came down our chimney and drank our offerings of a glass of port (it somehow managed to be two glasses in the end!) and ate the mince pie. Rudolph obviously wasn't hungry, however, and only took one bite out of his carrot. Typical isn't it, you try to offer an option of fruit and vegetables and look what happens?

It was the most lovely day, and indeed the roast potatoes were crunchy, the Yorkshire puddings did rise and the turkey was cooked fantastically well, if I do say so myself. Sebastian was like a dream, if a little bemused why all the family was around, everybody opening presents and acting just a little bit tipsy. I am sure he was even more bemused by the sight of both nannies asleep on the sofa snoring at one point.

I was also very lucky with my presents and my obvious love for bees and beekeeping has become well known. Jo bought me a lovely bee mug in which I have been having copious cups of tea today. I was hoping for a complete beehive, smoker, bee suit, hive tool, bee brush, solar wax extractor, honey extractor and every other beekeeping contraption known to man. However, Father Christmas was either not aware of such contraptions, was too worried about weighing down his sleigh or he simply couldn't bring himself to get the elves to build such things. Imagine having to go to the elves and state 'James Dearsley from Surrey in England would like a solar wax extractor'. There would be uproar about such a silly piece of equipment and why anyone in their right mind would want such a thing. Oh well, maybe next year!

I go to bed a very happy man. Primarily because it was my son's first proper Christmas but also because I have a wonderful family, a wonderful home and life couldn't be much better.

DECEMBER 31

Our celebrations on New Year's Eve are always a favourite of mine: celebrating the last year and looking forward to the next. Jo and I also now have a tradition to have a get-together with a group of our closest friends and this year was no different. Suffice to say we always have a lovely evening with the nicest of people; we eat far too much, play silly quizzes – which I always end up losing horrendously – have fabulous and interesting discussions, watch the fireworks and celebrations on TV, all before retiring to have a few more glasses of plonk or something a little bit stronger.

Another traditional aspect of our New Year's Eve celebrations is the challenges. I can't remember how it came about but several years ago we decided to make ourselves better people that following year.

I set myself a challenge for last year of making a Heston Blumenthal recipe. With Michelin stars aplenty, he is known to be one of the best chefs in the world with recipes such as snail porridge and mustard ice cream. Having seen a recent programme on the remaking of classic British food in his particular style and then realising there was an accompanying cookbook, I decided that I could do one of them. Earlier this month I realised that I had still yet to complete the challenge and so, for some reason I decided to be the cook on New Year's Eve using Heston's cookbook for inspiration to create a slightly different take on chilli con carne. It took me three days to make and cost me a fortune!

The pièce de résistance was using dry ice to make a sour cream sorbet to go with the chilli and, oh my, what fun I had with that. We were all there, champagne in hand while I put the sour cream into a mixer and then added dry ice. It was chaos. On went the blender and we were instantly covered in the most amazing smoke, the type you see at rock concerts. But within about thirty seconds I had sour cream sorbet and we were all walking on dry ice smoke clouds. Such a great experience and one I would recommend to anybody.

The conversation came around to our challenges for the coming year. My mind had obviously already been made up and there was no impulse predicting this year for me as there was last. Some of those around the table were obviously panicking. Matt wanted to start an apple orchard, Neil wanted to act in a Shakespearean play, Jill wanted to watch more silent movies from the 1930s and Jo wanted to learn Italian.

My turn came, and there was silence around the table. Considering they had just seen me almost break every implement in the kitchen with dry ice in a last ditch attempt to conclude last year's challenge, it was a big moment. It was the moment I had feared: the first time I would admit to the outside world that I was about to become a beekeeper.

'I would like to become a beekeeper and, with the bees' help, make one jar of honey next year.'

There was silence.

Matt eventually went, 'Wow.'

'Really?' said Jill.

'Fantastic!' said Neil.

Jo, whose head was in her hands, started to look up, visibly relieved that we weren't about to be ostracised by some of our closest friends.

My coming out as a beekeeper had gone well. We proceeded to talk about beekeeping and bees, the troubles they were in and how it affected us, not to mention how honey was made and how I would get a jar next year. Several glasses of port and wine later, we were merrily concocting stories of Italian silent films depicting Hamlet planting an apple orchard and putting beehives around the outside to help pollination. I was happy and I felt I was on my way.

Happy New Year.

JANUARY 1, 2010

Ouch.

For the first time I realised that a young child and New Year's Eve do not really work. Despite our going to bed at about 3 a.m. still laughing about Hamlet planting orchards, Sebastian was up at 6.59 a.m. just like clockwork, wanting a feed.

JANUARY 6

After the overindulgence of New Year's Eve I felt terrible for days, but today is my birthday. I am the grand age of thirty-one and feel a very lucky man. I have a wonderful wife and a lovely little man. I live in a wonderful area of the British Isles and have a nice job that, though it is hard work, stressful and I work long hours, affords me the life I would like to lead. Things do not get too much better but to top it all, Jo got me an amazing present. It is a book entitled *The Beekeeping Bible* and it is 2 feet thick with everything you could ever imagine needing to know about beekeeping inside. Part of Jo's inscription reads: 'Here are a few tips to get your one jar of honey this year.' A few tips??!! It will take me a year just to read the book.

JANUARY 15

A couple of months ago I decided I would not only chat with beekeepers in the flesh but also with beekeepers in the online community and it seems to be paying off. Originally I felt this would be a little bit weird. For starters I didn't know anything about Facebook, let alone beekeeping, but I am a quick learner.

I now have more beekeeping 'friends' in the US and Australia, let alone Greece, Turkey, Georgia and Bermuda than I do here in the UK. It seems that most of them have hundreds of hives. Here I am just starting out and aiming to get two! I feel a little bit silly talking to these experienced beekeepers about beekeeping when they seem to know so much.

As many of you will know, men are the true multitaskers and as I type this I am also having a 'live chat' with a beekeeper from Egypt called Mustafa. I still can't really get my head around it if I am honest and as he types another little message about 'queen rearing' I log off. What the hell is that about? I don't feel at all qualified to answer those sorts of questions.

Regardless of this latest experience, overall it has been positive so I decided today to join Twitter – a social media platform on the Internet that seems to be getting a lot of publicity at the moment. I had a little look around and there seem to be a lot of beekeepers on there. Let's see what I can learn in 140 characters (the maximum number of characters you can use to write a 'tweet'!). It all sounds a little silly to me but I am now @surreybeekeeper. I really don't know about this... Bring on the spring.

JANUARY 29

I have so far had a fantastic three months learning all about beekeeping but I am beginning to feel a little frustrated. I really want to get some hands-on experience, if anything just to look inside a hive. How will I feel when I open the roof for the first time – will I be scared? Will I have the nervous excitement of a four-year-old child or will I just be my relaxed self? Ultimately, will I get stung?

I want to know how it feels to have the bees flying all around you. I want to know how it feels to put on the bee suit (I wonder if it makes you feel invincible or will it just make you feel self-conscious and silly?). I want to remember that I must put elastic bands over my gloves/sleeves to stop bees wandering up them. I want to understand what it must feel like to have a veil on, let alone knowing how I will feel seeing a bee walk across my eyeline, millimetres from my nose. I want to know what the smell is like when you light the smoker for the first time. I want to know if I am brave enough to actually try to pick up a bee, to handle it without hurting it, just to see what they are all about. I want to pick a frame out of the hive as if I've been doing it all my life and check both sides in that sweeping movement that beekeepers make while checking for problems or looking for the queen (apparently if she is present and laying that is generally a sign of good health in the hive). I want to know if I will be able to find the queen when I am looking at thousands upon thousands of other bees...

Pause for breath.

I just want to start my journey and communicate my feelings and adventures to other aspiring beekeepers. I want to show people that, if I can become a beekeeper, anyone can.

Ultimately, at the end of the day, I just want to start making my one pot of honey.

JANUARY 30

Jo, Sebastian and I went to a National Trust garden in Esher. It was one of those lovely sunny winter days with a layer of frost covering the ground, which meant it was also bitterly cold! The gardens were beautiful and dated back to 1715 and as usual old Capability Brown had an input somewhere along the line, with the rolling hills and sporadically placed tree copses dotted around to make it look all natural.

We went back via RHS Wisley as I had been meaning to drop in there for some time now. The National Trust and the Royal Horticultural Society are two institutions I cannot fault. RHS Wisley has a fantastic library there with every gardening book you could conceivably imagine. I had a feeling that they may have a good selection of books about beekeeping and fortunately I was not disappointed.

Therefore on top of the house-sized *Beekeeping Bible* which I am trying desperately to get to grips with, I picked up *Keeping Bees: A Complete Practical Guide* by Paul Peacock. It is by far the most modern book I have seen and has the best pictures. Not sure if this makes for a great read but it looks far more inviting than some of the others.

I also picked up *A World Without Bees* by Alison Benjamin and Brian McCallum, which I am looking forward to reading as its opening fact states: 'If the bee disappeared off the surface of the globe then man would only have four years of life left.' This is supposedly a quote by Albert Einstein, though I have heard that it was a beekeeper by the name of Albert N. Stein, from the US – if this is indeed the truth, isn't it funny what Chinese whispers can do. Anyhow this book attempts to substantiate this claim and gives reasons for the problems they are facing.

I am also going to go through *Beekeeping: Self-Sufficiency* by Joanna Ryde, which also looks pretty new and quite 'fashionable' with its muted, earthy-coloured front cover and modern typeface. We shall see how well it reads! Everything else seems so old and textbook-like.

There I was in an RHS library, sitting on the floor with books all around me. There was a 'Yes' pile and a 'No' pile and then a sporadic jumble of books in a 'Don't Know' pile. I must have looked a bit of a test case and so when a lady approached me, I had a feeling she was about to ask me to leave for creating such a mess. However, she simply asked me if I needed any help with the beekeeping books.

As I looked up at this rather demure lady, books everywhere around me, I could hardly say no. It turned out that she was one of our regional bee inspectors. Her name was Dianne Steele and I couldn't believe my luck. Maybe she was off-duty but, either way, my preconceived vision of an inspector was not really coming true. There was no uniform, no medals of service and no yellow and black beret. Dianne was just normal, lovely and thrilled to speak about bees.

What a great person to bump into! Anyway she recommended *Bees at the Bottom of the Garden* by Alan Campion. To me, everything that *Keeping Bees* has in design, this book makes up for in functionality. It really doesn't look inspiring and is looking a little tired of life, but it does look like it is filled with detail.

Therefore on my pile of rather modern beekeeping books I also had one rather moth-eaten book, but I was hardly going to disagree with a bee inspector. I trudged off to the counter a little bit worried about just how much reading I was about to take on.

FEBRUARY 1

In just the short time I have been on Facebook and Twitter I now have over 500 beekeeping 'friends', and every day I get updates on what they are doing. It's very weird having all of these virtual friends. And then, on this beginners' page, I now have over 1,000 beekeepers following the updates – amazing really, in such a short time.

I am still working out exactly what I am doing on Facebook but it has certainly accelerated my learning alongside the course and books. Primarily it's because around the world, every beekeeper is at a different stage given their different climates, which gives me a fascinating insight. It is great to see photos and videos being shared of apiaries varying from one hive to several hundred hives. It's amazing to see the variety of hives and techniques as well as beekeepers.

I am learning so much from beekeepers called Chuck and Chad, not to mention the number of Vladimirs and Machmels I am now in contact with. I like to think it will make me a more rounded beekeeper.

FEBRUARY 7

Jo and I have just staggered back from a fantastic wedding in Oxford. Not only was it at Blenheim Palace, the birthplace of Winston Churchill, which was stunning, but it was, far more importantly and of far more historical importance, the first night we have been away from Sebastian together. It was quite surreal to actually have time together. We even got excited going into Beaconsfield Service Station, for example, knowing that we could sit down, have a coffee and read the paper! It is odd to have reached the stage where even a motorway service station is an exciting prospect.

However, I got into her bad books as I decided to extend my trial of whiskey. I have never liked whiskey but really want to 'learn' how to drink it. I went through three or four samples at way past bedtime. I just love the ideology of whiskey, the heavy tumbler, the ice, a roaring fire and a traditional drink that is steeped in history. The pub had all of those elements and so I couldn't escape it, especially as our friends Ian and Darren are seasoned whiskey drinkers. A couple of hours later, and feeling as if I could say I was a whiskey drinker, I stumbled away from the bar. Great wedding, Sarah and Ben, congratulations.

Sebastian loved being away from us, by the way. He had a great time with Nanny, not realising we had gone away.

On a separate note, I have now had more time to read through the plans for the beehive. They look far more complicated than I initially thought and this is not because they are somewhat misshapen and littered with teeth marks from Sebastian's attempts to eat them a few weeks back. I thought it was essentially a box. I hadn't really considered the types of joint that were required, the glue that was to be used or the paint to preserve the hive. It was all very particular and precise.

A hive should be perfectly square and have absolutely no holes or gaps anywhere if you want your bees to survive through the winter and generally accept the hive. It does state quite clearly 'competent woodworker required', as I noticed when I first opened the plans, and the fact that I struggle to identify what is wood and what isn't probably doesn't put me in this category. I was beginning to think that the Dearsley Bodge Job and a beehive were looking an unlikely combination.

There was another funny comment in the plans and my eyes just went out on stalks when I thought of the consequences! It states that the National hive is preferred by many because 'more hives may be packed on commercial vehicles or the domestic car' – yes, the domestic car! Who would be crazy enough to put a beehive in their own car? Can you imagine if just one of the little insects got out and the carnage that could cause, let alone all 60,000. I then read on a little further: '... and up to eight, transported in estate models.' WHAT!!!! One hive was mad; eight would be bonkers! Imagine if they all got out: 480,000 bees flying around the car. What on earth would other drivers think?

Tomorrow I will be sending my plans to my father to get prepared. I have to find the cedar wood to start with as this is what is recommended. Where on earth am I going to find that?

FEBRUARY 11

There I was, at work, logging on to my Twitter account as usual. I use it a lot in my corporate world, much to the amusement of my colleagues who hear me 'tweeting' all day long with the bird type sounds the computer makes every five minutes or so! I was settling down to my cup of coffee, catching up on the news and information flying past my eyes on the computer screen. Suddenly a message popped up from @stevefreeman, someone I had been conversing with and a relatively local beekeeper.

It read: 'Might have some bee news for you James, can you send me your email if you are still interested in a nuc or two? Thanks, S'.

Wow, could this be the first step to getting some bees? A 'nuc', or nucleus, is a small amount of bees; usually a hive will have eleven frames and a nuc will contain five, ideal for a beginner. Would my journey actually turn into something real? What exactly did he mean by 'bee news'? My heart was beating a little faster and I got a little excited. I realised this could be the moment my beekeeping life kicked off and became real.

I immediately sent a message back with my email address, trying to sound all nonchalant about it. Deep down though, it felt similar to being a teenager again after a date. Working out how keen you want to sound by the speed of your reply and then, when you do reply, carefully sculpting that reply to not sound too desperate.

An hour later and Steve hadn't replied. Maybe I had scared him off by replying about five seconds after the message. Oh no, my teenage years were coming back to haunt me. Perhaps I should have left it at least ten seconds before replying. It was awful; I was literally on the edge of my seat, waiting for a message.

Come the end of the day, there was still nothing. I sent Steve another message. Again, nothing. It is now 10 p.m. and I have just sent another. I'm slightly concerned now that I may be stalking him – another flashback from years past. I never thought I would be a beekeeper, let alone a stalking one. This is awful, I must stop.

FEBRUARY 13

It has been two days and I still haven't heard anything! Not even one miserly hello! Have the bees disappeared? Has he not picked up my 'tweet'? I feel like I have taken one step up the ladder of being a beekeeper only to have fallen off and landed on my backside.

I have to get in contact with him somehow. I mentioned it to a local beekeeper that I have got to know, Adam, who incidentally came to my attention by commenting on my blog and later we found out that he will probably be teaching me the practical side of beekeeping with the Reigate Beekeepers: a nice coincidence. He said that gaining a nuc of bees from people you don't know can be a dangerous thing to do because you won't know their history or what type of honeybee they are.

What type of honeybee? I thought it was quite simple. There were honeybees, bumblebees, and solitary bees. I now find out there are hundreds of different varieties within this set-up. Note to self: must read up on this pretty quickly!

FEBRUARY 14

As a tall, spotty youth, frustrated by the fact that my mum still cut my hair at the age of fourteen, resulting in a rather embarrassing kiss-curl in the middle of my forehead, Valentine's Day was always a tense occasion. There was always a nervous wait as the postman walked down the path and I would secretly watch him each year. This did stop after the year I got overexcited at a card actually being delivered to me which wasn't from my Nan. I opened it in a fit of joy as only a teenager could who had never before received a real Valentine's Day card. Imagine my additional excitement when it wasn't signed with a question mark but a name. I could work out the word 'Tiger' in a scrawl which temporarily got me even more excited until I realised that this was the name my best mate gave to his three-year-old sister. I made a pact never to look for a postman on Valentine's Day ever again.

I will admit to a little twinge of excitement yesterday though as the post dropped through the letterbox. I saw a large A4 envelope with my name on it and the small postmark of the BBKA (British Beekeepers' Association) but have only just had the chance to actually open it and peer in. I applied for membership a few weeks ago now and so this must mean that they have been mad enough to actually accept me as a member. The welcome pack included the following:

- A standard welcome letter
- A really useful booklet entitled 'Advice for Beekeepers'
- A bit about the organisational structure of the BBKA and its democratic notions (!)
- A leaflet called 'Bee Books, New and Old'
- A small leaflet entitled 'BBKA Enterprises'; basically what you can buy through them
- A leaflet stating that you could get some more leaflets from them about all the diseases
- A raffle ticket for me to purchase (must remember to do this)
- A FERA (Food and Environment Research Agency) booklet about 'managing varroa'
- My first copy of the monthly BBKA News

Now I have to say that it was lovely to receive this through the post but, if I am completely honest, I feel a little let down by it. Now that may sound harsh, but I mean it from a constructive perspective. It is obviously a very historic organisation (founded in 1874, it states on the paperwork) but the impression I get from this information is that it is still run by those historic methods.

The indicators of this include: photocopied sheets of paper, advertising leaflets that you can send off for (they do state it is available on the website but only small and at the bottom) and the *BBKA News*, which is simply words and no pictures. The information is fantastic but it isn't particularly inviting.

Now, we are all aware of the explosion in beekeeping at the moment. Surely this is the time to update methods and publications. I am sure that the new member coming forward will be like me, a slightly younger demographic (based on the evidence so far it stands true) and one that is not yet experienced in the art of beekeeping. Therefore, the *BBKA News* especially needs to be more inviting and readable, and ultimately needs to look after the new beekeepers that won't have the faintest clue what they are doing.

I am thrilled to be learning how to be a beekeeper and I am sure there are many others out there who are as well, but let's bring it up to date and introduce it to a new generation. It is always better to start at the grass roots, isn't it?

FEBRUARY 15

I am sitting here in the study, looking outside and feeling a little bit guilty.

This weekend was 'cut the willow' weekend. We ~~have~~ had a huge, beautiful willow dominating our garden. It was beginning to look like Sideshow Bob from *The Simpsons* and needed a haircut. My guilt is due to the fact that honeybees (and other foragers for that matter) love willow for early season nectar, something I never knew before. I didn't even realise that trees were a really important part of the honeybee's diet! My only justification was that I haven't any bees yet so next year they will love some new, fresh growth.

I had got some quotes before Christmas from local tree surgeons which were extortionate for what I saw as a small job: reduce the willow by a third. Looking back on the day, I now realise why they charge so much money. In my infinite wisdom I instead called on Bob, my next-door neighbour, a legend of a man who all men strive to be like. Every weekend he is outside with his chainsaw cutting up wood and alternating every so often with his gigantic axe. His downtime is spent fixing his equally humongous Land Rover with a V58 engine or maybe just tinkering with his runabout tractor (he doesn't own a field but just loves tractors). He goes to the pub for a Sunday afternoon pint at 4 p.m. every week and is the genuine article: a 'bloke'.

He very kindly offered to help me out as he knew a bit about trees and so he would do the difficult bit – climbing it and chopping it back – and I would do the easy bit, the clearing up down below. It all started well, though Bob started to get 'wobbles' three-quarters of the way up, which put me at ease a bit while doing the real job of raking up the fallen branches.

Anyway we made good progress and got most of the job done before Bob could go on no more and it was getting dark. He did work ridiculously hard. So 90 per cent of the tree is now cut. If the bees are quick, they might still get something. I put some money behind the bar at our local pub for Bob's 4 p.m. pint by way of a thank you and now I just have to finish all the tidying up.

Willow is not the easiest thing to tidy up. As soon as it sees a bit of open, bare skin, it decides to whip you – invariably across your cheeks. Not exactly a pleasant experience especially as there is so much of it. I hope I survive the clear-up without too much pain.

Bob, chainsaw in one hand, other manly equipment in the other, sloped off to the pub and a nice open fire. I gently put my rake in the shed, sat in its doorway, looked up at the stump of tree and, while having a cup of tea, contemplated the honeybee and the fact that I had just said goodbye to their spring forage. Sorry, bees.

It was only then that I saw Jo carrying Sebastian out of the front door, woolly hat firmly in place as it wasn't exactly warm outside. As she turned to walk towards me sitting there with my cup of tea, I could see Sebastian wildly gesticulating towards the front gate. It became apparent given the very evident noises of a chainsaw that he was making, along with his frantic waving as he approached the gate, that it was Bob he was after as he walked down the road in the general direction of the pub. Fathers with rakes just don't cut the mustard against neighbours with chainsaws, even to toddlers. Having my last, slightly cold sip of tea, I went to join them bidding Bob farewell.

FEBRUARY 16

It was a proud moment for me today. I returned home from work and as I opened the door to the house I could see Jo and Sebastian waiting for me at the top of the stairs. Jo had a lovely smile on her face and just said, 'Go on Sebastian, what noise does a bee make?' Sebastian looked up at Jo a little bemused as if not sure what was being asked of him. You could then see a little light-bulb moment as he turned to face me. A little discernible 'buzz, buzz, buzz' came out of his mouth and his face lit up with a gigantic smile. Funny, isn't it, how small things can make a parent so proud; here was my sixteen-month-old son making a buzzing noise and no matter how bad a day I might have had I now felt on top of the world. I rushed up the stairs to give him a great big slobbery kiss and a longer tickle than normal.

Something else happened today: Steve Freeman wrote back to me, and I now feel incredibly guilty for hounding him. The reason he wasn't getting back to me was because his eldest child was in hospital and it's also the reason why he wants to give up his bees. Despite the fact there was no way of me knowing this, I still feel terrible and I have to take my foot out of my mouth. I apologised as I think Steve took my enthusiasm (and looking back, frustration) the wrong way. My naivety has shown through like a beacon of light especially in the online world of Twitter where all your comments can be seen by everybody and 140 characters doesn't give you a lot of time to explain your comments! I don't suppose messages like 'Hello Steve, are you there? I am dying to hear from you' or 'Hello Steve, have you disappeared?' are too rude but when you put them in context with what he is going through, it probably wasn't the most sensitive of things to write. I think I can rule out getting bees from him now as I could tell from the text he had written that I had really annoyed him. I didn't dare ask him if I could still have his bees.

Even so I go to bed a very proud father.

FEBRUARY 18

One of the lessons I was taught during the beekeeping course was to speak to your local beekeepers. At some point, because of the fact that your bees will be flying up to 3 miles to collect food, there will be the inevitable mid-air collisions, turf wars and bees getting jealous of each other's queens being prettier than theirs. Perhaps, more seriously, speaking to local beekeepers is out of politeness, communication and support networks. However, the most important reason is to know the local issues that may affect your bees and what their bees are doing – whether there are any local diseases, what the honey flow is like – and simply to avoid any surprises. I might also ask them to tell me about the local crops to avoid another field-spotting road trip.

I know of two beekeepers near me who I feel I should make contact with. Both are probably within half a mile of where I am planning to put my bees. One lives in a fantastic house in the heart of the village, apparently runs seven hives and sells his honey to villagers. The other beekeeper is situated at the fabulous farm shop at the bottom of our small, bump-ridden, car-killing road. He runs several hives at the shop and sells the honey there. I originally thought it was the lovely lady at the shop who always takes a shine to Sebastian who made the honey, but when I asked she informed me that it was in fact a commercial beekeeper. Yikes! Was this guy really going to want to talk to me? He was obviously a very serious, experienced and commercial beekeeper; a far cry from my humble beginnings with no hive yet to speak of. She gave me his number to call.

It's funny but I felt quite nervous about trying to get hold of them as I simply didn't know what to expect. Here was I, this young, enthusiastic, naive, new guy trying to say hello (wasn't sure what else to say!) to these super-human beekeepers who have probably been doing it for years.

Anyway, this afternoon I spent some time drafting a nice, polite letter to the owner of the big house in the village, complimenting him on the fact that I noticed he has a nice garden (I couldn't really think what else to say) and just saying hello. I felt writing a letter was more polite than just turning up on their doorstep, so we will see. I popped it in the postbox. Derek was the name given to me by the lady at the farm shop and, as I had no address for him, I felt I should probably call him. Tentatively I tried this evening, not really knowing what to expect. The phone rang and a polite but firm voice answered, and so I responded. 'Hi Derek, my name is James,' I said, 'and I would like to have some hives near the farm in Newdigate, please.'

Though I nervously blurted it out, he was very nice about it all but admittedly he was worried at first that I just had this crazy idea to start beekeeping. This isn't too far from the truth, but he was relieved to hear that I had done a course and was obviously serious about it all. I suppose if a local beekeeper is inexperienced or doesn't tend his or her bees well it affects everyone locally so I can understand his reaction. After the initial awkwardness and inquisition, it was quite enlightening to speak to a beekeeper that had thirty hives, and it once again makes it all feel a little bit more real. He invited me to see his hives when he checks them in April – how exciting.

FEBRUARY 21

I have become used to sitting in front of a roaring fire, drinking either red wine or home-made cider, to write my diary. Tonight, however, it is being rudely interrupted by an airport lounge in Newcastle. Corporate life has taken me away up north where it is very very cold.

Having been stuck here for a couple of hours now, this post is being written while I wait in a lovely, colourful, inspiring departure lounge, not a drop of alcohol in sight, full of happy people filled with smiles from ear to ear – can you tell the hint of sarcasm here? We have just been told by a rather large man in a fluorescent jacket that they 'are currently assessing the runway to see if it is safe to fly due to the snowy conditions'! Everyone has their heads stuck firmly into books or devices which look suspiciously like iPhones (wish I had one), or devices that wish they were iPhones.

Due to my current situation, I thought I would just reflect on something that happened yesterday and today.

(Just been told we can board – hurray!!)

Yesterday was the day to finish the final 10 per cent of the willow haircut (or massacre). Bob, the man mountain, joined me in the afternoon and we got on with the job at hand. My God, I had forgotten how hard it is to keep on bending over and picking up willow.

(OK, well, I have boarded, am sitting on a seat which, being 6 foot 5 inches, means my knees are by my ears and the computer is somewhere under my nose, and have just been told that with the snow coming down we may not be able to fly... Ho hum.)

Anyway, so the man mountain and I finished the job of cutting the willow. I have to say it is lovely to end up with some fantastic willow poles which I shall make use of this year. But it was also a reminder that the gardening year hasn't really started yet. This is where it all started to go wrong and I found it quite tough. I haven't had time to tidy anything up; last year's geraniums are still looking dead in their pots – I must remember to bring them in next year to overwinter; the grass is all a little bit uneven, wanting a cut and needing to be rolled flat once more. I can't tell you how sad it is to walk on waterlogged grass.

Then, while I was in Gatwick Airport this cold, damp, typical February morning milling around WH Smith, I looked in the garden magazine section and in amongst them all was the friendly face of Alan Titchmarsh advertising an exclusive magazine, *The Gardener's Year*. I had a quick thumb through it, looking at some lovely pictures and my good mood was restored. They had some appealing photos of what gardens will be looking like in just a month or so, plus step-by-step plans to reassure you that the waterlogged grass, messy-looking pots and weed-filled beds are all normal, and instructions for what you should do about them. I felt the day had taken an abrupt turn for the better, knowing that it wouldn't be too long until the daffodils started to open, and after them the tulips, and after them God only knows.

If spring is coming, it must mean the bees will be too. It's getting closer – but I must now work out where to get my bees from.

FEBRUARY 22

I feel a little strange: I have just ordered my first hive. It feels a little bit unreal, perhaps because it was so simple. I had a good look around and felt from the information and pictures supplied by Nicholas at Peak-hives.co.uk that they were the best people to go with. I got onto the website, bish, bash, bosh, one hive ordered, which will make its way down here to Surrey from the foothills of the Peak District.

What a moment.

I decided, in the end, to opt for the slightly cheaper hive using red deal instead of the preferred cedar wood. I suppose the only real difference, aside from the price, is weight (deal being a little heavier I think) and the fact that cedar probably looks a little bit nicer with far fewer knots in the wood, which despite a few coats of paint will still show through.

So I now have to go and buy the frames to go inside the hive and then, hey presto, my first hive will be born! Whoopee!

Something else quite bizarre happened to me today. One of my Twitter contacts sent me a direct message about my question about which hive to get. @conchdraig, or Trevor as he is known in the real world, told me rather randomly to watch out for bears, deer, racoons and skunks when placing my hive. Gladly I soon ascertained that he is over in the States and so Trevor was justifiably more concerned about such predators than I am. I believe we only really have to watch out for the green woodpecker. A far cry from the troubles they have to go to! It looks like they literally build bee enclosures surrounded by electric fences or barbed wire to keep them all out. Thank the lordy that I may only have to consider some chicken wire around the hive in winter time if I have the slightest concern.

FEBRUARY 24

It's coming up to midnight and I have just got settled in front of the computer after an evening of corporate life. I feel exhausted and it couldn't have been further removed from my evening yesterday spent sowing sweet peas and chilli peppers. Having spent hours talking and incentivising I feel quite tired now and ready for a rather cold bed (as Jo is staying at the mother-in-law's tonight, along with Sebastian). I cannot wait for my head to hit the pillow though; it's funny, when you are tired, the words just don't want to come out particularly fluently but I just really want to get my thoughts down before the night-time displaces them elsewhere and the dawn chorus wakes me with other thoughts...

A lot of people seem to be taking an interest in my beekeeping exploits and my quest for just one pot of honey. There is such a nice community surrounding bees and beekeeping that it is lovely to say that I am becoming a part of it. Everyone is so keen to offer advice that it can actually be quite confusing, but I have to remember that they are doing it out of kindness and their passion for the bees. I get the feeling beekeeping is really about trying to understand all the information you can and then making your own conclusion about what feels right. It's good to know there is support at hand, though.

I was obviously excited about having bought my first hive, and mentioned it to a few of my beekeeping friends. There then followed a barrage of questions which included the following:

What floor are you using?

Did you get the crown board?

Have you thought about what foundation you are using?

Which frame type will you be using?

Are you using a stand?

All of a sudden there is a barrage of other, very important information to take on board and study. How did I know there were different hive floors and foundations? Why do I need a stand? To use the analogy of buying a car, I have basically picked the car body shape but nothing else. I now have to pick all the seats, the radio and the gear stick; not to mention the engine (let's say this would be the bees – and that is a whole different ball game!).

Adam, one of my regular online helpers, really showed me what a few years' experience will give you: the ability to reel off all the technical information, which I have only read about in books and still have a complete lack of understanding about how it all comes together! I look forward to the day that I can speak with Adam's authority.

I am continually getting asked the question about what bees I am going to get. I really must look into this as I am still only really aware that they are yellow and black. I am vaguely aware that some are from Africa and some are from Europe but I have no idea which bees are local to me around here; it doesn't help that I have not heard anything from the beekeeper in the village that I am sure could answer most of my questions. Is there an English bee or even better, a Surrey bee? I am quite looking forward to finding that out. I am also learning a lot about solitary bees at the moment, and the importance they also have on pollination, which has been interesting. Apparently they are a hundred times more prolific at pollinating than honeybees but, as their name suggests, they work on their own.

I must go to bed... These thoughts are not really making too much sense and another long day awaits me tomorrow.

FEBRUARY 27

It has been an interesting couple of days. First, there was the compelling event of our lovely neighbours moving out, who hadn't started packing until the day before. Imagine trying to empty a four bedroom house, with no professional help, while trying to look after two kids aged ten and three, two dogs, three cats, three chickens and two gerbils. It was never going to be a relaxing twenty-four hours for them and it certainly didn't look like a relaxing final hour either. As the removals company for our new neighbours, Nicky and Jo, arrived and just parked patiently outside, poor Duncan and Jane were literally throwing their lifelong belongings into the back of a small horsebox – yes, a small horsebox. Heaven knows why they left it so late but I should imagine their respective blood pressures have hit new highs.

We went to see them this morning in their caravan, which is only just big enough for them, let alone their menagerie. I also got to see the task ahead of them – building a four bedroom house from scratch. Their predicament looks pretty bleak, and when you consider that they couldn't fit much of their belongings inside the caravan it looks even worse. Last night they left it all outside under a loosely fitted tarpaulin, just in time for a pretty terrible rainstorm. Off flew the tarpaulin and needless to say most of their things are now sodden and unusable. It is not often I feel sorry for people, but I have to say this morning was one of those occasions as I stood in the abyss of a huge hole ready for foundations, which resembled a swimming pool after last night's downpour. I walked away pretty pleased with my own domestic arrangements but wishing them all the luck in the world.

Quite aside from all of that it was quite an exciting day for me as I saw an advert in the back of *Beecraft*, the beekeeping magazine, for a nucleus of bees for sale. After closer investigation

I realised it was for five frames of bees (about 5,000 apparently) complete with a 'laying queen'. The price of £150 is quite high really if you consider what you are getting for that money. But then as beekeeping seems to have become the 'in thing', I suppose they can charge that sort of money.

I was having some discussions with a blogging friend of mine about this sum of money and basically the crux is whether I definitely want to start beekeeping this year. The cheapest way to get bees is to hand your hive (or parts of it anyway) to your local association and then wait. You hope that as swarms of bees are recovered from around the local area, your hive is the one that is picked. I wonder if it is like those games you play on Brighton Pier which are a complete rip-off – you know, the ones with the mechanical arm that looks like it will pick up the fluffy bunny, nips its ear, starts to pull the bunny up and then, just as the pincers close, the ear miraculously escapes... There is no guarantee this way that you will get a swarm. However, there are a lot of beekeepers in the association and so there must be a chance that a beekeeper forgets to check their hive or doesn't see the tell-tale sign of a hive wanting to swarm (large cells known as queen cells are what to look for apparently). When the old queen gets the hint that the colony is raising a new queen, she will take a good proportion of the older bees and fly out of the hive to find a new home elsewhere. Generally this will be local to the hive for a period of time while they try to find a viable new home. If a swarm is reported there are people within local associations that will go and retrieve these swarms offering the bees a new hive to take up home in.

I was told of a little poem that beekeepers are said to remember when dealing with swarms. It goes like this:

A swarm of bees in May is worth a load of hay;
A swarm of bees in June is worth a silver spoon;
A swarm of bees in July isn't worth a fly

Essentially the earlier in the year you can get a swarm the better. If you can get an early swarm, then there is more chance that they will become established and you could get a good crop of honey from them.

Alternatively, if you buy a nucleus, you will have bees, simple as that.

I have also learned that having a nucleus, despite the expense, is also good for the beginner. As the frames of bees grow, so does your experience and confidence. If you are lucky, you may get some honey at the end of the year as well.

I think that I may try both. I might just have to make contact with this gentleman, to see how many nuclei he has.

There is one more option. You can put an empty hive near your house and hope that there might be a passing swarm that may be vaguely attracted to the colour, aspect, size and smell of the lovely empty hive that has appeared on the horizon. It sounds a complete shot in the dark but could be worth a try as, except for the effort of taking a hive and planting it on the ground somewhere, it doesn't sound like there is a lot to do. There are ways to increase your chances of attracting a swarm to a bait hive, the name given to a hive for this purpose. You can rub propolis into it – a glue-like product produced by bees – or rub other bee-friendly smells into internal walls like orange or lemon peel.

Either way I have to pick one of these methods and make a decision. No bees mean no honey!

The last bit of exciting news is that my local beekeeping association is having a 'frame-building' day next week – which, as it happens, is the same weekend that I will be building my first hive (hopefully I receive it in time through the post). A year ago I would have laughed at the chances of going to a frame-building day... What have I become?!

Finally, today I was able to get out into the garden and plant some shallots and onions as well as do a general clear up. It felt really nice to be in the garden again and the weather held off, which was a bonus. I had a really enjoyable day with Jo and Sebastian, running around the shops and then going to a playground in the afternoon. It's great to actually have time as a family; a rare commodity at the moment due to work commitments and other friends and family commitments. We do too much! Tomorrow: working, driving up to Birmingham, on to Bolton and then home again. Not the nicest way to spend a Sunday.

FEBRUARY 28

Though it is a little bit late to do this, today was an important day. I felt Sebastian was old enough to learn how to chit potatoes, even if it meant doing this from his high chair. On reflection, I am not quite sure he has yet grasped the importance of letting the potato sprout before planting and still sees the potato as an object to throw or nibble on.

In my family the potato is the most important of all vegetables and I have fond memories of my formative years helping Dad on the allotment to dig them up. It was always as if Dad was as amazed as I was every time the fork lifted up and lots of lovely, dirty-looking potatoes came spilling out of the earth. I remember distinctly Dad's love for pink fur apple potatoes and he and I would always marvel at the bizarre and silly shapes that would come out of the ground.

In thirty years' time I wonder if Sebastian will have the same nostalgic feelings for the vegetable that I do. I will do my best to ensure that he does and it all started today while trying to describe to him why we place potatoes into half-sided egg cartons to let little sprouts develop. I am not sure he was too impressed but I think he felt it was a great break from his usual shape-sorters where he attempts to put brightly coloured squares into circle-shaped holes. This was a breeze compared to that and despite looking distinctly unimpressed by his first attempts at potato chitting he was actually rather good at it. Now there is a job for life in my book.

MARCH 1

Spring is officially on its way. Goodbye February, hello March. I have to say it was one of those days when you draw the curtains and the first thing you notice is that fella Jack Frost has made an appearance. In the half-light you can make out a hazy sun rising as well as seeing the wildlife around you waking up to a beautiful dawn.

It is generally accepted that Britain is a pretty miserable and inhospitable place when the weather is foul, and it has been for about three months now. However, today was an exception. It was stunning and it's amazing what a little bit of sunshine does for you. Today, Monday, was great for many reasons and it all started with the weather.

I live in the middle of nowhere and good God do I know it. Having lived in London for many years, the move to the country was a welcome relief; but the most fundamental change, alongside the saddening realisation that I couldn't walk out of my door to a choice of pubs and restaurants, was the fact that I really noticed sunrise and sunset. I cannot tell you how much I despise the clocks going back in the winter when the commute to and from work, which by that time is already in semi-darkness, is plunged into pitch black. We don't have streetlights out here so you really, really notice the dark. However, one of my favourite times of the year is fast approaching and it is simply gauged by the sunset I witness. Each day, driving home at pretty much the same time, days just seem to last that little bit longer and though it's only about two or three minutes a day, it's enough to lift my spirits an incredible amount. Today was one of those days; the first day I have noticed the last strip of light as I was driving home. How wonderful.

Arriving home, I turned the key in the lock and pushed the door open, and standing there at the top of the stairs was Sebastian. At the top of his voice he simply shouted 'Dad-dy!' Now I have got the odd spattering of 'Dad-dy' before but not with such annunciation and sheer joy. What a lovely coming home present from my seventeen-month-old son. He then proceeded to run off and cause havoc with his fire engine but, by that point, I wasn't really worried.

This has also been a nice day because of the dawning realisation that this is the week that I could get my hive and could be building it at the weekend – whether the hive arrives or not, I will be building frames for a beehive, courtesy of my local beekeeping association. Though a little apprehensive, as it will be my first meeting with them, I am looking forward to getting my hands dirty.

There is one final positive point about today, though, and that is about the old beekeeping neighbour of ours, Anne Buckingham whose Saab I used to clean every week for extra pocket money; my parents must have bumped in to her while walking the dogs and Mum phoned this evening to tell me. Her garden was beautiful with a great slope covered in flowers and she had kept her beehives high up on the outer perimeter. It may explain, on a subconscious level, why I felt drawn to beekeeping. I always remember the yellow spots of pollen we used to get on the washing when it was hung outside not to mention the spots on the car which were impossible to remove. Mum always came in and stated it was because of the bees down the road.

Now, what I didn't realise was that Anne was, and still is, very active in the Surrey Beekeepers' Association. What a wonderful coincidence. I have asked Mum to see if she would mind me popping over for a cup of tea and a little bit of advice to steer me in the right direction. Maybe this weekend after I have built the hive... unless I lose my temper while doing the fiddly bits, which I am sure could happen.

What a nice day; wish they were all like this.

MARCH 3

For every good day like Monday, there are some days you are just glad to get through, and today is one of those days. It was all going well until the final thirty minutes before I was planning to leave work and I spoke to one of the staff over in Spain (I work for a company that works mainly in the UK and Spain). He stated that more people were being made redundant today and others were taking a pay cut, some for the second time. We must have seen over 75 per cent of our workforce go now. This is a tough recession; this sort of thing obviously gets you thinking.

As a result of this I missed my first bee association evening meeting due to consoling some of my Spanish colleagues and reassuring some of my English ones. The insecurity these situations bring is incredible. Thankfully, I feel that is where hobbies come in handy, keeping you focussed and as a result keeping you sane. This morning as dawn broke I went up to water my seeds and saw the little green heads of the broad beans poking out along with a couple of sweet peas. Seeing these little signs of life each year always gives me a boost. A sign of the year to come. I cannot wait to use my recently acquired willow poles to train the sweet peas up.

In the past I have been guilty of being a workaholic, but I can feel a tide of change. These are tough times. Not only the worst recession for a generation, but also coming out of the worst winter in a generation, tends to change your outlook on things. For me it is a time to reassess my work–life balance. It has started with gardening and will continue with beekeeping.

MARCH 6

I am feeling quite overwhelmed today after an amazing day. After the negatives of Wednesday where I missed my first bee meeting, today I did manage to attend a frame-building day, which involved about thirty new, slightly rough around the edges, theoretical beekeepers keen to hone their practical skills. I am sure it is simply a way to get new members building frames for the old members!

It entailed a nice man giving a demonstration on using the basic elements of woodwork to put together a frame for the bees to lay the honeycomb on. Basically you need some wood, which looked like balsa wood, and eleven small nails which hold it all together. You also need strips of wax as well, which already has some hexagon shapes to encourage the bees (though without the wax they would still make exactly the same shape and size!). It felt a little like *The Generation Game* as he was able to build this frame in a little over four minutes and then said 'Over to you'! There we were, all keen and eager to compete for the cuddly toy, gathering the wood, nails and hammers to have a go.

I did the first one in about ten minutes and felt pretty confident about it but then, on the second, realised that one vital tool was missing from my repertoire – pliers! I blithely nailed the wrong section and then got in a complete mess. I desperately needed pliers to pull out the wrongly placed and protruding nails so I could start again. As I was looking at the mess, I realised that I had also managed to put my hammer right through the wax and therefore left a gaping hole in the middle. Not all was lost however, as apparently the bees would fill in the gap with their own design. Amazing; but sadly I didn't think I would get any further in my quest for the cuddly toy. The other contestants just carried on unperturbed.

It was quite a chilly morning, and they have been for the last week despite fantastically glorious sunshine throughout the day, but you could see a few bees flying around. The site for this frame-building day was just outside the Reigate Beekeepers' apiary, a fenced-off area containing about ten hives. Every time the sun made an appearance, I spotted a few of them flying down to areas with water – I suppose they were gathering some up to take back to the hive – a lovely sight.

Towards the end of the frame-building it became evident that something I was wholly unprepared for was about to happen. It was suggested that we might actually pop into the apiary and see if the bees were all right after their long winter of cuddling up together to keep warm. I was completely taken aback as I thought today only my woodworking skills would be tested, not my comfort levels, being surrounded by hundreds if not thousands of bees. On reflection I am quite glad that I didn't know beforehand as I would have been quite nervous.

Anyway, in for a penny, in for a pound... I saw most of the others had gone to get their bee suits. By that time there was only a small and medium left and they were all jackets, not the full-on body armour protection I had expected to be wearing. Being a not-so-slim 6 foot 5 inches tall, this wouldn't be the best start but everyone was already getting going. Richard, someone I had met at the training course back before Christmas, helped me with the veil which felt distinctly weird to have around my face, and then I was ready to go.

I looked a complete idiot wearing the medium-sized suit and I felt slightly uncomfortable given the amount of flesh on show, especially the large expanse of back that was left exposed by my suit riding up. I had heard that bees love to land and then walk upwards into dark areas – that was all I needed with a sizeable square footage of rump back skin available for them. I was also wearing jeans which I have heard is a complete no-no when it comes to bees as they just don't seem to like them. Let me assure you that bees do not insist on a dress code, at least I hope they don't, but I hear that the fabric of the jeans is not great for the legs of the bees and they can easily get caught which is not good for either party.

I was expecting to be given some gloves to protect my digits from attack but all I was given was some of those membrane-thin ones akin to what surgeons wear. I couldn't quite believe that these would give me any protection and my hands felt very exposed, but these give you far more control than thick leather gloves, which apparently is far better for the bees.

If truth be told, I felt a little like a teenager given a jumper knitted by their gran for Christmas that was based on a design for a five-year-old, but too polite to take it off.

I was put in the same group as Richard, and we were introduced to our mentors, Tom and Maggie. As we were marched in like ants, I quickly took Maggie to one side, who looked less likely to judge than Tom, and subtly suggested that it was like taking a lamb to the slaughter with me dressed in the suit. Maggie, fresh with enthusiasm and bursting with energy, just told me that it would be OK and that, to be honest, it was usually the head and eyes that they went for. She rejoined the group as I tried to digest this last little nugget of information. At least I had a veil so I slowly, rather nervously trudged into the line and wandered into the apiary.

The first thing that struck me as I entered the apiary was how small it was. There were ten hives in total, meaning that in the height of summer they would house over half a million bees in an area probably about 25 feet by about 20 feet – can you imagine if they all swarmed at once! It seemed almost claustrophobically small given the content within.

Before I could go much further with this train of thought, Tom went right up to the first hive and took the roof off. Underneath the roof was the cover board, the small wooden membrane that separates the world from the bees. There was no time to be nervous but my mouth was getting slightly dry at this point. Would there be many bees? Would they fly all around me? Would they make a beeline for me? As I went into a trance-like state, Tom kindly asked me to move away from the entrance as I would be right in their way as they flew into and out of the hive. Needless to say I moved out of the way very quickly. Tom then lifted the cover board a fraction to puff some smoke from the newly lit smoker inside. There were a few bees crawling around. Then after a few minutes he lifted the cover board off completely and started to remove the frames. I watched, fascinated.

Tom was checking the level of stores available on each frame; apparently this was necessary as we might need to feed them until they could start to regularly forage again. As he removed a frame to check it, hundreds of bees would fly into the air. It was an incredible sensation to see thirty beekeepers surrounded by thousands of bees that were coming out of the hives all around us and the sound was immense. The buzzing was incredible, and powerful is perhaps the only way I can describe it. Sitting in the garden on a summer's day and hearing a wasp or bee fly by or nestle into the flower next to you is nothing like the sound of thousands of bees flying around you. I suppose I have just come to terms with opening up one beehive and being surrounded by some bees. Seeing this multiplied by so much for my first time was both mesmerising and terrifying. I was struggling to keep calm and put aside the urge to make a run for it. For all of my thirty-one years I have become very apt at avoiding situations with bees or wasps as I was taught that anything looking vaguely yellow and black is going to hurt you. Here I was surrounded by them and as time went on I started to feel all right about it. I think it helped that everyone around me was in the same situation.

There were a couple of times when the bees landed on my hands which felt a little too close for comfort. I suppose, on reflection, it is similar to having a small spider crawling over your hand with that ever so slight tickling sensation as they wander around. Like with the spider, it is only the visual aspect of it that makes the tickling sensation more apparent and, in my case I was very thankful when they decided to fly off elsewhere. Throughout the whole session I was convinced that I could feel something crawling up my back, which can only be because of my rather small bee suit and the mental tricks it was playing on my mind. I must have looked quite strange constantly pulling down my bee suit over my back. Another strange sensation was seeing a bee crawl right in front of me, across the veil, millimetres from my nose. I felt like that cartoon dog when the bee lands on its nose and its eyes go crossed keeping sight of it. I have not crossed my eyes since I was about seven years old, which was only a matter of fun and I was trying to impress a girl. Here I was, as an adult, staring cross-eyed at a completely different female form and feeling equally stupid in the process.

The twenty minutes or so we spent inside the apiary, although uncomfortable at times, were utterly compelling and really captivated me. A lot of what I had learned on the course fell into place at that point. Suddenly I could understand the differences between the hives; we went through a WBC, a National and a polystyrene one – apparently these always get the bees started early in the year as they manage to keep the bees warmer in winter, but are a nightmare to clean, though they are a lot cheaper. It really helped me understand the fact that the bees all cuddle together to keep warm, to learn about the honey stores and where they are, about the capped egg cells, what a queen looks like... I could keep going and going.

Once I had struggled out of my bee suit, relieved I hadn't been stung – in fact nobody had – we went through the crucial skill of lighting a smoker. If it goes out mid-inspection I can only imagine the chaos as the smoker is there to calm the bees down. It makes them feel that the hive is on fire and so they concentrate on gathering stores to leave the hive rather than attack whoever is entering their hive. After a while they realise that the smoke has gone and they go back to normal. However, if a beekeeper is still checking through a hive I would imagine it to be quite fun and the bees may get a little feisty. This is why it was a really useful skill to learn and one I will be pleased to get under my belt. As far as I am concerned, the smoker is as much a part of beekeeping as the hive and the suit. It is almost synonymous with the hobby and so I was glad to get the chance to light one. I thought it would be easy but alas, no.

First of all you have to pack the smoker with paper and get it lit. This is all fine and relatively easy as long as the match stays alight. The fire can get going pretty quickly and is soon burning outside of the small stainless steel smoker. Then you have the skill of adding a fuel to the smoker to keep it burning for a period of time. The trick, however, was not putting in too much in order to keep the fire alight, but enough to emit this beautiful smoke which would be important for the hive inspection. Several people put in too much which extinguished the fire immediately despite desperate attempts to get the fire going. Others would only put in a small amount and never get it smoking. It was a fine balance getting it right and a real pleasure when you had done so.

It is a wonderful contraption with bellows doubling up as handles which, when squeezed, force air into the chambers and puff smoke out of the funnel at the top. Whoever came up with the design deserves a medal. To get the small fire going we used cola nut shells, which smelled amazing and apparently when burned emit a 'cool smoke'. I never knew there was a difference personally – I thought smoke was smoke! Apparently a great variety of substances can be burned to make the smoke. These include cardboard and dry twigs, but these emit a hotter smoke and can even send out sparks as you squeeze the bellows. Not great I would think for the bees' well-being or temperament!

My turn came and despite keen eyes analysing every move I made I was fortunate to get it lit first time and the smoke quickly followed with a few rather too enthusiastic puffs. A couple of burning embers did fly out but no damage was done. It feels so stupid to write, but I felt both proud and also strangely masculine. Man must have fire, and all that! I have to say though, it wasn't as easy as it looks and I would say only about 50 per cent of us got the smokers lit first time but it was great fun trying.

Sadly after a couple of hours I had to leave as they were all still there honing their new-found skills with enthusiasm. There is one thing that I really took away from the day: it is absolutely crucial to join an association. The fee I paid Reigate Beekeepers' on an annual basis was worth it just for that session. There were about thirty new beekeepers there and I would guess about seven more that were experienced beekeepers on hand to help out, answer questions and make tea and coffee (how fantastic that it was made over a Calor gas makeshift stove with a giant, almost witch-like, kettle – always makes it taste better!). There cannot be too many hobbies or pastimes which you can learn with that degree of help from more experienced practitioners.

In summary, I learned a lot about bees. I learned that despite thirty years of running away from them, I could actually stand in an area surrounded by them. Ultimately, I learned that beekeeping really is a way of life, that people are truly passionate about the little things and are so keen to help others realise this same feeling. What a truly unique morning.

MARCH 7

This weekend was about Dad and I building a beehive from scratch, which sadly in the end was never to materialise. When Dad and I looked over the plans the bodge was never going to work. We were both very concerned at the level of detail and the fact that your measurements had to be exact and angles perfect. This is not something we are particularly good at.

Instead of building a hive I decided to pop next door with Dad and see Anne, my parents' neighbour, and her bees. Fortunately Anne was outside and she recognised me immediately.

She was as lovely as I remembered and we had a great talk about her bees. She is currently running five hives, a mixture of the beautiful WBC hives and Nationals, which I could see in the distance in the same elevated position I remembered from childhood.

Anne insisted we go on a tour and I gladly followed, with a newly brewed mug of tea in hand. We approached her hives with none of the protection of yesterday's association experience. It was therefore rather nerve-wracking, and I felt my heart thumping as we walked within 3 feet of the hives. Positioning ourselves behind them, we had a lovely view of the garden but also a stunning view of the bees flying in and out of the little entrances at the front of the hive. With the fine weather it was obviously warm enough to start foraging and it was fun to see them starting their journeys or coming in loaded up with pollen.

Anne was so confident in her movements, even lifting off a roof at one point to peer inside; I didn't dare look and kept back. She invited me to approach the side of a hive and I tentatively agreed. Moving as slowly and carefully as I could, I crouched down to watch the bees landing before entering the hive. I was not even a couple of metres from the bees and without a suit but it was amazing seeing these bees with pollen loaded on their back legs landing on the landing board and walking up into the hive. I got to fully appreciate why people enjoy sitting by the hives just watching the bees coming and going. It was incredibly relaxing to watch.

Anne and I strolled back across the garden, and despite her giving me a bit of a Spanish Inquisition about my motivations for starting, it was only when I discussed the courses and the reading that I think she realised how serious I was about doing it properly. I do feel that more experienced beekeepers are very concerned at the wave of new beekeepers coming into the hobby and that they start it all up correctly. This is the second time I have felt like I have had an interview from a more experienced beekeeper and it is like they are extremely protective of the bees much as a parent is of a child. It cannot be a bad thing.

To finish off our little tour, Anne then showed me her bee shed where she stocked all her equipment. It was perfect. A proper old wooden shed, well-used and characterful – the incredible sight was inside though. It was filled, floor to ceiling, with beehive parts. I was now able to identify most parts so could see brood boxes and supers everywhere, not to mention frames which pretty much filled every tiny gap left. With several bees flying around for good measure it really was the perfect bee shed. I must get one of these!

Regretfully I had to get back home for lunch but it was lovely to see Anne again and I received lots of useful advice, including the possible need for four supers per hive (the supers are situated just above the brood box where the queen lays her eggs, and are where the bees deposit their honey), whereas I had only ordered two. Most of all it was just nice to see yet another friendly face willing to help me along on my journey.

MARCH 8

I received a funny tweet from @kezdiman, who was describing his friend who is based in Transylvania and is obviously a large-scale beekeeper. It kind of expanded on a theme from @cochraig earlier but really brings home the differences we all face:

> '100+ hives in the Vrancea mountain wilderness. Armed with 3 dogs, an electric fence, he is on a 24-hour vigil 4 bears!'

My only issue will be green woodpeckers – apparently they might get hungry in winter, peck a hole in the side of a hive and raid it for honey – pales in comparison!

MARCH 9

I have created a blog, I have a Facebook and a Twitter page, I have commented on forum sites about being a beginner beekeeper and my flat-packed hive arrives on Friday – hurrah! But there is still one slight problem: how am I going to get some bees? Are my efforts all a pipe dream? I thought I would have my bees by now. I really should have got this organised earlier.

One of my options is to buy a nucleus of bees from another beekeeper or company, and so I rooted around for the advert I had seen ages ago, as that particular beekeeper wasn't too far away. I hoped he hadn't sold all of his nuclei as that would put me back again.

I had done a little bit of research on the beekeeper by asking others whether they knew him and they all said he was very reputable (I would hate to think what a non-reputable beekeeper would be like) and so I made contact. Yesterday it was confirmed that I could get some bees from him but would be waiting till the end of April or beginning of May before I could get my hands on them as he said that the colonies are taking a little bit of time to get going. He didn't want to give me a small colony so suggested I just wait a little longer.

Either way it will leave me with one hive having a nucleus of bees and I will then hopefully get a swarm in the other. Mmm. I hope this will allow enough time to get that jar of honey from at least one of them!

But at least I have now ordered some bees.

MARCH 13

On reflection, today has been a great day. It did start early, at 6 a.m. – this time of year I always like to get something done early in the morning before Jo and I wake Sebastian up and I always run out of time what with all the seed sowing and general preparation work for the spring – and I planned to pop up to the allotment to get the first sowing of potatoes and mangetout into the ground. When the mornings are clear and sunny like today, it is simply beautiful and a joy to be up early. It is quiet, no one is around and you can go about your business without real time constraints as essentially you have borrowed time; it feels more peaceful knowing that you are ahead of everyone.

After the initial success in planting the spuds in almost straight rows, I moved on to the near-impossible task of preparing the remaining raised beds in preparation for yet more potatoes. With almost impenetrable clay as the topsoil, I can assure you it isn't particularly fun, and a visit to the chiropractor may now be in order.

I went and had breakfast with Jo and Sebastian. I love Saturdays as I can actually spend some time with them over breakfast rather than clock-watching. Sitting there in a tight-fitting suit waiting for the dreaded time when it is all systems go is never something I look forward to. There then follows the cold walk to the car, traffic jams, road rage all around me and the rolling up of sleeves as I walk into the office to face the first of many meetings and coffees. Did I mention the hours stuck in front of a computer? Needless to say I would rather stay at home!

Once breakfast was cleared away and Sebastian had been put down for a nap, I went back out. My job was to move the fantastic collection of willow poles I collected while cutting back the willow from our driveway up to the allotment. It felt like such a satisfying job to finally get rid of it from the drive – and as a bonus, serious brownie points earned with 'she who must be obeyed'. As I was offloading my second carload, however, my morning took a turn. Over the brow of the hill came a collie dog. This meant one thing: any second now, the farmer who owned the field would arrive.

Now let me explain the origin of my allotment. Two years ago I dropped a handwritten note into the bungalow at the end of the farm track that we live on. The recipient would be our rather eccentric and scary farmer known as Ray who I had never talked to but had heard lots about. I was requesting a little bit of his land to build my allotment on. Fortunately a week later I received a scribbled note back through the letterbox giving me permission to do so. We agreed a great-sized plot, about the size of a tennis court, which looked perfect and the rest is history; I have been breaking my back ever since trying to make the heavy clay soil workable.

Farmer Ray is an interesting character. His family have been working this land for centuries. He is passionate about chickens (he used to be an egg farmer) and is genuinely someone I would like to get to know but not like to upset. I would guess he is in his mid sixties and though he scares me half to death, he has this aura of being a man of principle and I would certainly invite him to the pub for a pint; for his stories about The Beatles alone.

Let me digress a little bit. When I first met Farmer Ray to discuss which bit of his field I could borrow he started to tell me about his run-ins with The Beatles back in the 1960s. Forgetting my quest momentarily, my ears pricked up to what seemed a bizarre but fascinating story. He started to tell me how they used to drive past his farm on the way to a little shack by the pond nearby for all-night parties with the likes of Jane Asher. It sounded as if they didn't get on too well; he didn't like their convoy of blacked-out Minis and went about stopping them coming up the road, patrolling it with his shotgun.

Apparently he once parked his trailer outside the entrance to the shack, blocking them all in, and he left it there for the weekend 'to teach them a lesson'. This might give you a little insight into the guy I was now dealing with! There can't have been too many people who would have done this to The Beatles at the height of their powers.

He then started telling me how the lyrics in some of their songs were about him (in particular the song 'I Am the Walrus' which contains a line about being the egg man). I thought that quite plausible but wasn't so sure about his claim to the song 'Bungalow Bill' being about him. He does live in a bungalow but surely it should have been 'Bungalow Ray'? But I wasn't about to disagree with him.

Anyway, in past weeks I've been thinking about the possibility of putting my hives onto the allotment so, seeing his collie dog approaching, I thought it could be the perfect time to ask this pretty important question. Putting this into perspective, I have got a hive (arriving on Monday I hope, as it was a little delayed) and I have ordered some bees, not to mention that I have taken a course. My last major stumbling block would be putting them somewhere. If he said no I was pretty stuffed. I couldn't put them in the garden – it just isn't big enough and I couldn't see myself persuading Jo to allow me to put them there. To be honest, with Sebastian, I wouldn't feel comfortable with that anyway.

It was therefore a stroke of luck to meet Farmer Ray randomly like this, so I walked over to him (though it felt like I was on my knees shuffling towards him, much like Smithers does with Mr Burns in *The Simpsons*) and started to say hello in a rather jovial, nervous manner, paying particular notice to his dog rather than him.

After some small talk I just blurted out something like, 'So, I was thinking, can I have some beehives on your field? I have done a course and everything.' Silence.

'Yes, of course. I love honey and I know bees are in trouble. Any honey you have spare would be appreciated.'

I felt like kissing him and then quickly remembered that this was Farmer Ray and quickly retracted that thought. Three months of concern had just gone out of the window as I now had a destination for my hives.

He bid me farewell and off he went, hobbling across his field (unploughed for 125 years apparently), man's best friend at his side. He had a hip replacement only six weeks ago. Hard as nails is Farmer Ray.

What a lovely man. He has allowed me to grow an allotment on his land and now he is allowing me to have some beehives. I felt elated and realised I must be on his good side.

MARCH 18

Last night as I was driving home from work, Jo called. 'DHL must have arrived while we were out and delivered a package for you. It's so big and heavy I have just left it outside.'

My heart did actually, believe it or not, skip a little. So my hive had arrived. It was too late to do anything about it as it was pitch black outside, but another day was not going to hurt. I did pop outside later on, though, to see a giant black shadow of a box. It was far bigger than I had expected and I went back inside very excited indeed.

I got up early this morning to attend to the small jobs in the garden that I don't want to be doing at the weekend. I walked around to where DHL had hidden the package. Rather embarrassingly, they had left it behind last year's Christmas tree that I had been meaning to burn for months now! Though my feelings from last night were confirmed about its size, it was considerably lighter than I had expected.

There were in fact two boxes and so I took them around to the front of the house and started to open them excitedly. I felt a slight pang of nostalgia as it was a feeling akin to when Father Christmas used to deliver my presents on Christmas Day.

So I started to open it up and laughed, as this beautiful hive had been protected by bits of recycled cardboard evidently from children's toys. The first piece I pulled out was from 'The Little Princess' and the second was 'Pots and Pans'. This only added to the sense of Christmas nostalgia knowing that, due to timing, these were probably from Christmas presents given to the hive-maker's children. It was also further evidence that this was bought from a small business as it had all the touches of someone actually hand-making it. I loved the fact that the nails provided with the hive were in a recycled envelope with 'Nails' scrawled across the front. How wonderful and non-corporate.

I took the gabled roof out of the box and it looked lovely. I could have gone for the flat roof but felt I needed to upgrade to this sloping roof, akin to one you'd see on a house, for my first purchase. Basically it makes it look more traditional and I am so pleased I decided to go with this little bit of vanity as it really finishes it off.

There were so many pieces and yet no instructions. How on this earth was I, the person who was literally thrown out of woodworking class, going to put this together?! Having given up on the idea of building a hive from scratch, perhaps my pride had got the better of me when ordering. As with all hives, I had the option of either flat packed or assembled. The latter was about 10 per cent dearer so I opted for the former.

Regardless of the lack of instructions, I shall have a go this weekend with the said nails and glue for added reassurance and try not to make too much of a hash of things. Let's see how it goes!

MARCH 20

It has been a truly glorious morning and I got up early to review the task at hand. I felt ready to go but not before Jo and Sebastian got up and breakfast was completed, carnage as always with food splattered everywhere. Sebastian is getting to grips with eating on his own now, which is always an experience to watch. Now peace reigns once more as Jo has taken Sebastian out for a walk. It's a perfect time to get building.

As I was opening the package on the kitchen table again to get to grips with it, I was feeling quite excited and something that had not struck me previously was the smell of the wood. It was as if it had just been cut – a really delightful smell. I counted twenty-six separate segments of wood as I took them out of the box, all of which looked pretty similar. The only bit of woodwork I have ever done in my short life was building a bench similar to those you would see in any park around the world. You know, the ones you steer clear of because they are either inhabited by the local sleeping tramp or by slightly strange people sitting down on them, knees firmly together, balancing a clear plastic lunchbox while they bite down into a triangular sandwich with the crusts cut off. My bench wasn't exactly the greatest success as I measured up a little bit wrong and couldn't really work out how to drill holes properly. No tramp would have ever gone near it. I was therefore a little bit tentative about fitting together these twenty-six random bits of wood.

I really could have done with some instructions. Seeing there were so many different sections I felt I should have a dry run with no nails or glue and just put similar-sized parts together. With the aid merely of a picture of a National hive, it would be interesting to see what I could build.

I made a tentative start and felt my use of the set square was particularly good, having watched a YouTube clip on how to make sure you attain perfect symmetry. (I can't believe I've admitted watching a video of this nature!) In a surprisingly short space of time I had built the brood box, which is where the main nest of bees remains and the queen lays her eggs. I was feeling pretty pleased and the result wasn't bad. Who needs instructions? The two supers quickly followed suit and as a result, in front of me sat my first beehive – though if I even blew on it lightly, it would have all collapsed in a giant pile, and so next week I shall attempt the real build complete with nails and glue. I felt really chuffed and finished just in time for Jo and Sebastian to walk through the back door.

No word of a lie, Sebastian took one look at me, then looked at the hive, shook his head and promptly walked into the dining room. Jo followed, laughing, though I am pretty sure it was at Sebastian's reaction rather than the hive. Once we all had lunch and Sebastian had his midday nap we left for a short walk to the farm shop at the end of the road. Jo and I sat in glorious sunshine enjoying a cream tea, something you have to do when the sun is shining in the UK. We were enjoying watching Sebastian trying to run after the older children who were just that little bit more adept at running. Every five steps or so he would fall back onto his bottom followed by a great bout of laughter which would occasionally turn into belly laughs as he realised that the other children were laughing as well. It was glorious sitting out in the sunshine as we very rarely get the opportunity to do this sort of thing due to our busy schedules. However, with spring evident all around us with lambs jumping around in the grass and the daffodils and early tulips in their full glory it was sheer bliss.

I am feeling a very lucky man today as I also have to put in here another exciting event that happened today. Jo and I found out that we will be expecting baby number two in December. We have always planned to have two children but this really tops off a great day. I find it hard enough to comprehend being a father of one some days, but here I am considering being a father of two pretty soon. Well, rather more than considering, at this point. Can't quite believe it – but how lucky we are to have this opportunity once more.

MARCH 25

So, I went back to the original site where I bought the hive to look at the right glue to use, not to mention paint which I hear also has to be a certain kind, and there in front of me were the instructions telling you how to build it – not only instructions but also diagrams. It became perfectly clear I hadn't followed them and that was perhaps why my hive and theirs looked so different. I now realise my 'dry run' was a complete disaster.

As you may have fathomed by now, I am of the male variety. Yes, that's right, the type that would be happier to carry on driving into the wilderness rather than wind down the window and ask the man with the map standing by the side of the road for directions. To add to this I never, ever read instructions. As far as I am concerned the TV makes itself work with or without instructions. Therefore why should I waste valuable time reading a manual several inches thick when I could be enjoying my new purchase?

Despite seeing the dry run as a challenge tantamount to a complex jigsaw puzzle, I could see quite clearly from the photos that I had made a few fundamental errors – not least I had built the entire stand the wrong way round and the very important hand holds upside down.

I am going to review these instructions carefully before starting the real build so as not to make the same mistakes again. My corporate world is taking me away this weekend and so that is out of the question, the Easter weekend is looking most likely now.

Thank God I didn't just build it blindly like I would normally do. What would the bees have thought? What would *my son* have thought?

Note to self: I must buy my bee jacket and smoker. I can't imagine getting the bees and not being able to transfer them to this newly built hive. What on earth will I go for? Even just looking at bee jackets there seems to be a lot of different colours and makes to choose from so I better start doing some research.

MARCH 29

Feeling quite exhausted today due to a hectic three-day schedule at an exhibition in London for my corporate life. I had forgotten how tiring it is to stand up continually for twelve hours and just talk to people. To add to the intensity I was asked to be on a panel as an 'expert speaker'. How that happened I do not know! It was quite fun, however, and a good experience.

It was only on the Friday morning that I realised it was being held at Earls Court – such is my way of dealing with things very late in the day. When I got there I realised it was the Ideal Home Show at the same time. Being an exhibitor in another show hall meant I could gain free access to it. I knew that one of my Twitter bee contacts was there working on the Omlet stand, makers of the Beehaus. Having had quite a lot of correspondence it would be nice to say hi and a perfect opportunity to see the hive itself before possibly ordering one.

When I got the chance to pop in on Saturday evening, I have to say I was quite impressed. The description of it being a brightly coloured freezer box was spot on; the one in the show was almost fluorescent in colour. It did look better in reality, though, which was a relief. Another bonus seemed to be its ease of construction, and as there was no wood in sight there was no woodwork needed. Perfect! The Beehaus looked ideal for my second hive for comparison's sake, and I couldn't wait to order one.

A comedian recently commented that he knew he was getting old as he had actually met all of his friends. I know exactly what he meant now that I have all these online 'friends'. Saturday was the first day I physically met one of them and fortunately Quinn was as nice in person as he was on Twitter. It's always difficult to know what they are really like in little snapshot sentences of 140 characters but Quinn seemed a lovely guy and as camp as a row of pink tents, which only added to his charm. An asset to his company certainly, he further convinced me to purchase a Beehaus at some point but I simply cannot justify the price tag at the moment. Starting up beekeeping isn't exactly cheap and this is just one purchase too much for now.

APRIL 1

On reflection, I wish that I had sat back at the start of this year and made myself a list. After all, I am a bit of a list man. Literally every day I write lists for almost everything that I do. Unless I can tick things off throughout the day I am a little bit lost and that is pretty much where I am now. I seem to be obtaining equipment a little bit randomly and as and when I think about it.

Today I decided I had to buy a bee suit, and so did a great Internet search. It seems there are lots of options out there. I did a detailed search, looking at the colours, quality and price. I have to be honest, usually I will try to get away with the cheapest option but I wasn't that keen to go for the bargain basement option here. The last thing I wanted was to wear out a bee suit quickly and discover a hole mid-inspection which the bees would have discovered moments before.

Instead I went to the company that most people seemed to accept as the best suit company out there, B. J. Sherriff, and ordered my suit. Apart from the fact that most of my association wear their suits, I think it was more the fact that they have this great big 'Sherriff' badge as their logo on the breast of the suit so it reminds me of playing Cowboys and Indians as a child. I never got to be the sheriff as a boy and was constantly shot at by my mates and had my bows and arrows broken in two. Now was my chance!

As a result of the suit purchase, I also ordered a variety of other items, which included a rather large and shiny looking smoker and something called a hive tool, which I knew to be important.

Looking forward to when they arrive, if only to try on the bee suit.

APRIL 4

The package arrives.

Today I am feeling another step closer to my one pot of honey as my bee suit, smoker and other fancy, highly technological equipment arrived. Since we live in the middle of nowhere, our post usually arrives about a year behind everyone else's and so I was quite surprised it arrived so quickly. Though I had ordered an extra-large bee suit, large smoker and hive tool and was expecting a huge parcel I was amazed to see that it was in fact tiny; little more than Easter egg-sized, apt for the time of year but far more exciting!

I opened the box with anticipation, with Jo in tow wanting to see what on earth I was going to look like.

When I opened the parcel the first item that caught my eye was a fantastic shiny smoker. When I say shiny I mean mirror shiny. The extent of the shine brought back memories of my years as a mountain climber at university. Back then, shiny equipment was looked down upon. We would get to the mountain with all of our climbing gear wrapped around our waist and I remember the day that I had just bought my own gear and headed up to the base of the mountain with pride. I felt like a cowboy who walks into the wrong bar and the music stops, like I had just said something really inappropriate in the lull of conversation. Everyone was looking at me and I could see them looking at each other and I am sure I saw one whispering to the other, 'Oh look, here's a novice'; and they would all stop climbing and watch me attempt a climb. I remember so well the urge to get out each individual bit of kit and scratch it so no one would know next time. Every time I get something shiny I think of that moment and it always makes me smile.

Now I imagined reaching the apiary, and as others saw the sparkle of my new smoker they would gather round as I tried to light it, ramping up the pressure until the puffs of smoke came regularly, white and even – which is always a joyful moment, apparently.

Continuing through the package, next out was the 'hive tool', which is used to help you get out each frame when you are checking the bees. I have heard others refer to it as the 'beekeeper's friend' as you can't really do anything without it. I unravelled the tissue paper it was wrapped in and then all of the bubble wrap and out popped the brightest bit of kit I have ever seen in my life. It was bright orange and not only that, it was shiny as well. It is undoubtedly the brightest thing that I have ever owned. My thoughts, however, were then taken to a very different place when Jo mentioned what a lovely ice pick it would make. I am not sure if it is just me but a female voice uttering the words 'wonderful ice pick' transports me back to the film *Fatal Attraction* and instant nightmares. I quietly hid the hive tool in my jacket and moved quickly on.

Finally the moment of truth: the bee suit. I had opted for something quite untraditional. Firstly I had gone for a jacket style rather than full body, and secondly I had opted for a khaki colour rather than white. I had settled for a jacket as when I tried one on it was very comfortable, and after all it is all about what you feel comfortable in. I remembered people saying that full suits could get hot in summer and the thought of wearing a rather fetching boiler suit in the height of summer didn't really appeal, so I thought this might be better for me. The colour was a more personal choice but hearing the theory behind it was quite interesting. Apparently some years ago the founder of the B. J. Sherriff Company went to New Zealand to observe beekeeping there. They would do a lot of their beekeeping at night time and often found that white bee suits made themselves very obvious to the bees. Wearing khaki toned down the brightness and made it easier for the beekeepers to work. So not only was it a nice colour it also meant it would be easier for me to keep bees in the dark. It seemed the perfect combination in my eyes.

As I pulled it out, Jo peering over my shoulder, I was having second thoughts. Would the colour be OK? Did I pick the right size? Would Jo divorce me seeing what I was about to start wearing most weekends?

Staring at it through the cellophane packet it came in, it actually looked great and Jo immediately agreed but stated she would have preferred it in white because she's a traditionalist. It did look slightly too big but then I did order XXL. I think I was overcompensating for my medium episode earlier in the year and afraid that I might again feel the little feet of a bee on the small of my back. Not again, thank you very much.

I tried it on and looked in the mirror and felt pleased I had gone for the jacket style. Jo even tried it on as well though it looked more like a tent. Wearing this I don't think I could ever look like a traditional beekeeper; a fact I quite like to be honest.

All in all, it was a big moment and I feel one step closer to that one pot of honey – though I am starting to slightly panic about building the hive as I am putting it off every day. I keep on finding a new excuse. Let's make that an aim for this week!

APRIL 8

My local association holds practical sessions each week and I was fortunate to get to one last night. I think they are a fantastic incentive to join an association as it's a great way to (a) meet other beekeepers in a similar situation, and (b) learn the art of handling bees in a safe environment. I hate to think what mistakes I would make if I went at it on my own! This one was held at the local apiary, where the frame-building day was held earlier, and thankfully it wasn't raining, unbelievable given the last few weeks.

I drove there feeling a little more confident than I did at the frame-building day, but purely as a result of having a correctly fitting bee suit. I also brought the free gift I received from the bee suit company, some nicely fitting though rather unattractive-looking leather gloves. It certainly didn't look as though a sting would penetrate these little beauties (it didn't look like anything would penetrate them if I am honest!).

I got there in good time but I was shocked at the sheer number of people at the apiary already. There must have been fifty beekeepers, most of them in their white bee suits already. I have to say, I stuck out like a sore thumb in my khaki suit! Though the majority were in their fifties and sixties there was a whole group of teenagers present as well. It meant a wonderful mixture of ages but I was amazed as I wasn't the youngest present. I was only stating this point to a few friendly, fellow non-middle-aged beekeepers when whispers could be heard. These youngsters were Duke of Edinburgh participants. So these guys were effectively doing this to get a little certificate and not for any other madcap reason. Therefore, looking around I was now officially the youngest beekeeper there.

I was getting some looks from other, rather jealous, white-suited beekeepers; I think they were envious looks but I could be confusing envy with disbelief. We split off into groups of five and were shown the basics of a few fundamental skills by some more experienced beekeepers, in our case Tom and Maggie once again.

Let me tell you a little bit about Tom and Maggie. They will be my tutors for the next year; instrumental, therefore, if I am to get my jar of honey. Tom seemed about sixty, the more dominant of the pairing and the joker; obviously with an attitude young for his years. He had one of those faces a photographer would love, being friendly and yet characterful with deep features (and wrinkles) defining it. Maggie, delightful and a lot younger than Tom, seemed to be the perfect support act. She was the true powerhouse, however, and their approach epitomised my relationship with Jo: I wear the trousers but Jo tells me the colour to wear. You could see already that she was the brains to Tom's character. As he trailed off during a description or forgot something, Maggie was already chipping in. They are a perfect team and hopefully I will learn an awful lot from them.

Once our group's smoker was lit we made our way to one of the hives. On went the veil; I felt so smug that I had at last got a suit that fit me! On went the gloves and suddenly it all went silent in my group and everyone turned around to look at me. Tom and Maggie started to slowly shake their heads and I felt it must have been something to do with my gloves. I slowly surveyed the group's hands – something I haven't done when in the presence of a group of people before – and then realised that I was the odd one out. I was wearing gloves that looked like they would be more at home handling nuclear material.

The others were all wearing what we'd worn last time, what I would call 'doctors' gloves', those really thin, almost see-through, horrible things. What was wrong? I was surely wearing gloves to protect myself from anything and these guys were wearing essentially nothing! Tom politely suggested I swap gloves and said mine would be useless. I have to say, as soon as I saw them manipulating the bees I understood why. They were able to hold frames easily, pick bees up and generally do anything.

The idea, I found out talking to Maggie about it later on, is that the surgeons' gloves allow you to be more tactile and as a result the bees are calmer. With my oversized leather ones, if I wasn't careful, it could result in me being clumsier, the last thing you need when moving frames around. I thought that a fair enough argument.

The session continued and it was lovely to be around the bees again. I actually got the chance to remove a frame of bees and have a look for the queen, then place it back in the hive again – another small, though rather nerve-wracking, step forward. It sounds so simple but it involved the use of a hive tool to loosen the frame, then I had to try to pick up the frame (I see why Tom had suggested I change gloves) all the while fighting the urge to flick my hands if any bee landed on me. All said and done though, I felt a little bit more prepared this time and wasn't quite so scared. I even watched very closely as bees landed on my veil, right in front of my nose, and I was actually happy to watch them walk across it, millimetres from my face.

However, one small event started to turn my rather serene mood. Buying a jacket rather than full bee suit meant that my legs were open to the elements. I went for light-coloured jeans despite hearing that jeans weren't brilliant as bees' legs can get stuck in the fibres; unfortunately, jeans were pretty much all I had available. I watched them on my nice new bee suit quite unfazed and then I saw them on my trousers. This was the moment it dawned on me that these jeans were my gardening trousers. I wasn't overly concerned about the fact they were quite worn but it did occur to me that I had a 'button fly'...

... And there were a few bees taking an interest in that area. One in fact had its head in the fly already. She had free access to a very sensitive area and was already halfway inside. Panic set in, my heartbeat picked up and a long breath in was taken as I remembered bees like to investigate dark, warm areas. A button fly was like a landing strip to possibly the warmest area of my body, which would also be dark. My mind was racing. Here I was staring at a bee halfway into my button fly, working out how to get in further. I was imagining getting stung on the crown jewels. What on earth happens in that event!? Has anyone been stung down there before? Would it swell up? All of these seemingly absurd questions were rushing through my head as I watched a little yellow and black insect inspect the workmanship of my button fly.

I very quickly, and as subtly as possible so as not to draw too much attention to myself, made a flicking motion with my fantastically tactile and flexible gloves (thank God for Tom's suggestion as my leather ones wouldn't have stood a chance). I am proud to say that my first experience of 'handling a bee' was not the very cool-looking movement of pinching a bee carefully between thumb and forefinger as I had seen Tom do earlier but flicking one. Had I opted for the former I would more likely have given it a helping hand to discover what lay beyond my button fly.

Fortunately it was successful: the bee flew away and swollen private parts were averted! Needless to say the rest of the session was spent with my hands protecting that region while trying to look relatively normal amongst my new-found friends. Not an easy task, believe me.

Once the session was over, and I could start breathing normally again, I felt I needed a drink. As luck would have it, Tom approached the group and mentioned he was popping to the pub for a pint. Keen to get to know my mentor and some others, I agreed.

Of the new beekeepers, only three of us went along but as it turned out, most of the senior echelons of the association joined us. There I was surrounded by most of the mentors, not to mention the treasurer and the chairman. Overawed is perhaps a little strong but it was a little bit daunting joining a group who obviously knew each other very well. This is where the evening turned into more of an initiation ceremony.

Seeing some new beekeepers had joined them, old stories started coming out. It began rather innocently. Tales of honey collections flowered. Then it seemed to go off on a tangent to the subject of stings, as if testing our resolve and to see if we would turn up next week. It reached a peak when Tom started to talk in a whisper, leaning lower over his pint glass. We naive new beginners hung on to his words as he told us about a beekeeper who got stung on the eye. It was only after the descriptions of turning blue, foaming at the mouth and collapsing whilst writhing around in agony that I started to get the joke and called him on it. Everyone burst out laughing; but he still insisted he had seen someone get stung in the eye and that though his description was slightly exaggerated, it does happen. Bees automatically go for the face, he was saying, and hence being suited is always advised if near to the hive. Even so, as I took a long sip of my beer, I started to relive my very own encounter this evening, 'The Battle of the Button Fly', and I couldn't help but take the notion seriously.

All in all a great evening, some nice people met, valuable lessons learned and more knowledge gained.

APRIL 16

Today was the first day of the build. I have to get it done as quickly as possible due to the fact that corporate life is taking me away tomorrow, Wednesday, to Amsterdam for the day. On Thursday I then go away to Finland till Sunday. Arriving back from Finland at midnight, our week-long family holiday starts seven hours later on Monday morning. Looking forward, it doesn't give me a lot of time to build, paint and preserve the hive in time for the bees to arrive! Unfortunately, it is now midnight and so far I have built only the stand. The dry run obviously didn't help that much.

Bearing in mind my tendency to break either wood or my fingers while knocking in nails I am not holding out too much hope for the building process. I popped around to our new neighbours this evening as they are offering me a shed for free; how nice is that? I have to move it by Friday, however, and so was talking to them about how on earth I was going to dismantle it and then move it 300 yards to the allotment while out of the country. I think I have taken on a little bit too much.

It was therefore about 9 p.m. before I started and I took a while to make sure that the kitchen table was suitably covered with newspaper to stop any mess. The stand didn't take too long but the time was swallowed up by a moment of overconfidence resulting in my hammering a nail into completely the wrong piece of wood. A pair of pliers and forty-five minutes later, construction got back on track. How annoying. But I was feeling pretty pleased with myself that I had built something. I was feeling a little bit guilty, though, as I was still banging nails into wood at 11 p.m. I probably wasn't the neighbours' best friend!

All in all a successful night and now I just have to look at doing the brood boxes and supers which are essentially wooden boxes and so hopefully slightly easier – even though I messed up the dry run completely and put the handles on back to front and upside down.

APRIL 17

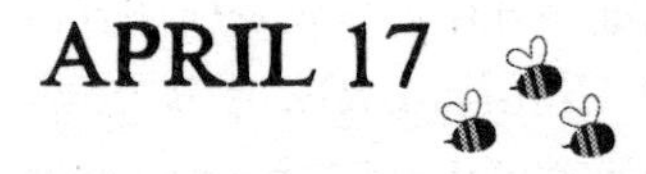

Back from Amsterdam, another tough day at work, and what better way to release the tension than to bang some nails into wood! On reflection I am pretty pleased with yesterday and so far the construction of the brood and super boxes has gone mysteriously well. I gained confidence, the nails seemed to be not only going in but also going in perfectly straight. Then, no sooner had I started, than I finished – and at a relatively comfortable 8 p.m. It was remarkable. I have to say; it made me wonder why on earth people buy complete hives when this was so easy.

However, I was still against a deadline as I am flying to Finland tomorrow, and so I decided to paint the hive straight away to at least get one coat of paint done before I go. My bees supposedly arrive at the end of the month and beekeepers recommend at least seven days' grace after painting. This allows the fumes given by the paint to subside. I am cutting it pretty fine at the moment – especially given I need to put on at least two coats, leaving twenty-four hours between each coat!

Painting is not my thing and I usually find an excuse to avoid it but I was quite happy to paint a hive as it felt somewhat different. There is also a knack to painting a hive as it is often better to paint all the outside at once. When you leave it to dry however, you have to be careful to avoid getting the floor covered in paint or getting the hive stuck to something. Therefore, it is best practice to build a scaffolding platform structure involving two stepladders and a pole between them on which you thread the hive body – it is much like a rotisserie allowing me to spin the hive around and will allow me to paint all sides and let them all dry at the same time.

With my obvious time constraints I opted for the scaffolding structure, despite it taking over the kitchen for an hour or two. I am not sure Jo will be too keen navigating a scaffolding structure to do anything in here so I will probably lose some brownie points. I really hope I can get this jar of honey to offer some form of payback.

Having completed the first coat of the brood box and super on the rotisserie it is now a waiting game until I can put on the second coat but fingers crossed the bees don't arrive too quickly as there will be nowhere for them to go.

A dawning thought looms as I write this. I have a lovely hive now but what happens when I get the bees as that is all I have – an empty box. Usually they need frames complete with wax foundation... Damn, none ordered yet. So I need to order some frames and foundation as soon as possible. Slight panic is setting in as I know they also come flat-packed and so I have to build them despite my mad travel schedule coming up. Oh dear, should have planned this a little better.

APRIL 18

Of all the things in the world, I never ever expected this to happen. A volcano in Iceland has erupted with such force that a huge plume of ash is now floating around in the atmosphere. This has grounded almost every plane in Europe, which means that I am not travelling to Finland after all. Talking to my staff around Europe it is like a disaster movie. Take the journey to Finland from Estonia for example; everyone is jumping on boats, panicking because flights are not taking place. It sounds like pandemonium! Bees wouldn't panic so much, I am sure. A little volcanic ash wouldn't put them off.

The good news, though, is that I will have time to add another coat of paint this evening. I had a little issue after last night's painting. When I went to bed last night, I thought the various parts of the hive were dry and so I stacked them up and put them away. When I came to pick them up today I realised they were all stuck together and it took me a while to split them apart. It was a full-on operation complete with the biggest kitchen knife I could find to penetrate the now dried Sadolin paint (apparently good for painting hives – clearly tough stuff!). If I was having this much trouble with the Sadolin, how much trouble would I have with the propolis? On second thoughts I should have used the hive tool here to prise apart the sections of hive but it completely slipped my mind.

Thankfully there was no damage done and the second coat went on with no problems. I never ever thought I would hear myself saying that I was able to paint my beehive because an Icelandic volcano erupted but life always throws up those little experiences.

What a wonderful coincidence and at least something good came of the event.

APRIL 19

The volcano just keeps erupting. The cloud of ash still seems to want to make its way toward Europe rather than in any other direction. Flights are grounded still and therefore one great positive, since we live near Gatwick, is that there is no aeroplane noise! To top it all we also have the most beautiful weather as well – so much for April showers. Today I have clear blue skies and complete silence. Bliss.

The second coat of paint on the beehive is looking good and therefore I am not sure I am going to do a third coat. Having had a minor panic about the frames and foundation the other day, I got on and tried to order them today. However, as ever, my timing was terrible. Anyone who is anyone in the beekeeping world is at the Spring Convention, near the National Beekeeping Centre situated in the Midlands. Basically, anyone learning the art of beekeeping should go as there are practical demonstrations, you can speak to the oracles of beekeeping first hand and basically be immersed in the topic. To top it all off, all the companies are there offering their wares cheaply. Why didn't I think of this before? Due to my trip to Finland I hadn't planned to go and so I didn't order tickets. Had I known that one of the most amazing natural disasters of modern times was about to occur I would have gone.

Therefore, when calling companies about sending me some frames and foundation they were all out, or packing and too busy to deal with my order. I left a number of rambling messages but I also bet that when they get back next week they will have run out of everything after the convention. I wish I had thought of this sooner. This is a real pain.

On a brighter note, and as a consequence of gaining a weekend – what a lovely feeling that is, incidentally – the need for a beekeeping purchase was felt. As a treat to myself I have got round to ordering that brightly coloured top-loading fridge-freezer of a hive, and I feel pretty pleased about it. I cannot wait to compare and contrast.

Will the new, fancy but expensive equipment trying to muscle its way into the marketplace be better than the older, more trusted equipment? It is quite funny in the beekeeping circles the cynicism given to the Omlet Beehaus. Beekeeping is a traditional pursuit and new-fangled technology is often viewed with scepticism. This is the same the world over if you consider it, however, whether you are talking about beehives or the invention of a new bit of technology in the car industry. I want to view it with fresh eyes because, after all, I have never used the traditional hive anyway, so I have no bias and therefore no expectation.

APRIL 20

Today I spent some fantastic time on the allotment and in the garden. The free shed I got from next door has finally been transported up to the allotment with the help of the neighbours. I just have to build a base for it now and then start to build the thing!

It is great to see the mangetout looking good and just starting their tangled journey up the pea sticks. The garlic plants are now about 10 inches high and the onions have finally started to get going. The Red Baron ones I planted only two weeks ago so they have some time to go. The shallots are just getting going now and the broad beans are thankfully no longer being eaten by pigeons. The pea sticks are acting as a fantastic deterrent for now.

As for the garden, the early daffs are now dying and the others are in full flower. The plum blossom is sporadic, but just days away from complete bloom. The apple tree is getting there with some sprouting leaves. I have seen one head starting to emerge from the *Alliums* but none of the others seem to be going anywhere. I hear some beekeepers get two collections of honey; the first being from early blossom like this. I wonder if you can taste the blossom.

Everything seems to be kicking into life. What a lovely time of year. I only hope that next year I will be able to appreciate the blossom while seeing my bees enjoying it too.

Off on holiday tomorrow till Friday so I am going to put my beekeeping thoughts on hold for five days and enjoy the family. I cannot wait.

APRIL 26

What a lovely week, our first family holiday with Sebastian talking and walking around, which made for its own enjoyment and challenges. We went to Center Parcs so he also got his first experience of being on a bike which he most certainly enjoyed – though it wasn't as fun for me as he kept walloping my back from his little kiddie seat attached to my own seat. It was as if I was his servant and not getting him around quick enough. The fact that after each wallop he burst out laughing made it worthwhile though and I was happy to oblige.

Despite promising Jo that I wouldn't work on holiday, I did press upon her the urgency of ordering my frames and foundation. I called a company recommended to me as having the best foundation, which was KBS. What I really liked was the fact that he was happy to sell me the foundation but advised me to buy my frames elsewhere. I love that honesty. Having now ordered both I feel a lot happier.

Arriving back from holiday this evening, I saw that both had landed on my doorstep so I am good to go; I just have to build them all now.

APRIL 28

It is late and I still have the slight smell of smoke around me from my latest trip to the apiary; that smoky smell has already become synonymous with my weekly trips to the apiary, quite distinctive and not at all unpleasant. Today I got the chance to actually lift out a frame from a hive and have a real look around all on my own. It's quite amazing to think that in a few weeks' time, once I have my own bees, this will become second nature. For the time being, however, I still get a little bit nervous.

It was a really nice session, with wonderful weather to enjoy the evening of hive investigation. All the groups split up and tend the same hive each week to see how it is getting on. Sadly our hive is not really doing very much and the poor queen is running the risk of being dethroned by her own daughters in favour of a new queen. She is simply not laying many eggs and so the hive is just not doing anything. Apparently the bees will pick up on this fact and want to replace her for the good of the colony. We shall wait with bated breath until next week...

MAY 1

After the most amazing April when we Brits were out sunbathing, making sure that lobster colour was on full show, the bank holiday weekend brought awful weather – just when my shed needs to be built! It always happens in the UK, just when it looks like we are going to have three nice days off the weathermen decide that this weekend will be terrible.

Having just bought a fantastic new hat for garden duties (the last one died after we thought the snowman's head was cold last winter, it was never the same again), I was ready to stroll out to the allotment with tools in hand to build the shed. The clouds looked a little gloomy but nothing was likely to happen. It takes all of two minutes for me to walk to the allotment and I felt especially good today knowing that it was highly likely I would have my Man Shed up by the end of play. Despite Jo calling it a Wendy House, thanks to its previous use as a children's playhouse when the old neighbours owned it, I have labeled it the Man Shed. In truth it will be where I store my beekeeping equipment, as its preferred location is next to where I think I will pop the beehives. I never thought I would have a shed to put all my beekeeping equipment in like Anne, my old beekeeping next-door neighbour, but here is my opportunity if only I could get a nice weekend.

As I opened the front door and stepped outside, the heavens opened. I literally mean horizontal rain, the sort only witnessed in the Amazon. I soldiered on, not to be put off, with the true British spirit of 'Don't worry, it's only a shower'. I arrived at the allotment and the foundations I had already laid, and tried to put up one side of the shed. It was all constructed and so was basically a case of putting nails into the four sides and roof. The first of the sides promptly fell back into the hedge and almost took me with it. Saturated, I gave in, acknowledging that I would need a bit of help here, and trudged back to the house in a strop. It was highly unlikely I would get the shed up at all due to the bad weather and lack of help. This was not good news as I needed somewhere to store things like extra frames and supers near the beehives for easy access. I am not sure I would be allowed to keep this sort of stuff in the house.

Back at home with a cup of tea to lift my spirits, I started to have a look at the bee frames. These needed to be built in the next week as I could get the call anytime to go and collect my bees. The first thing I realised was just how many little bits of wood there were! Only two hours ago I was getting angry at the four large parts of ridiculously heavy shed while being pelted with Amazonian rain. Here I was messing around with small fiddly bits of plywood and 15 or 20 millimetre nails (the size depends on where they go on the frame) which, quite frankly, I couldn't even hold, let alone hammer into a frame. Oh I am not in a good mood today.

A dry run got underway and with the exception of putting the wax in the wrong way, forgetting the order the wood is meant to fit together and forgetting that I actually needed to put nails in, I didn't think I did too badly. The wax was only slightly warped and the wood was only slightly out of place. I decided to do the rest this evening once Sebastian has gone to bed.

MAY 2

I must just offer some advice. If you want to survive to a ripe old age, don't build bee frames late on a Saturday night, watching football on TV, drinking beer while your wife is having a little kip on the sofa. My activities weren't well received. It was amazingly productive, though, and almost all of the frames were built and with only a few mistakes.

Very kindly and quite unexpectedly, my mum and dad came over this morning to help me with the Man Shed. The rain was again lashing down. We made our way to the allotment with all the kit and I have to say, it wasn't too bad in the end. The whole job must have taken about an hour and a half with only a few minor disagreements. I would say that about an hour and a quarter of that however was spent fixing the foundation that I had laid but despite all of this my allotment now had a Man Shed; a place I can escape to or at the very least store my beekeeping equipment.

I walked away from the allotment very wet but happy and very appreciative of the help I received from my folks. I wonder if in thirty years Jo and I will be doing the same thing for Sebastian?

MAY 4

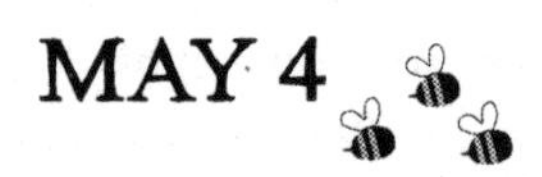

No news yet about the bees. I am hoping it won't be too long as I need to get a wriggle on if I am to get this jar of honey. It is looking less likely each day as I had been expecting a call or some contact by now.

I took the finished frames (complete with protruding nails, where I realised pretty early on that using 20 millimetre nails was the wrong thing to do in certain areas, and that was the reason that the instructions suggested 15 millimetres) up to the allotment. My beehive is currently situated behind the shed, facing south-east, which means it will get the sun pretty early on; very important, I am led to understand.

With the frames in the hive, I realised I had stupidly not made enough. I had put in ten and needed eleven. I also realised I was lacking yet another piece of equipment; the dummy board; quite apt really. This is half the size of a frame, and slips into supers and brood boxes and, once removed, allows you to manipulate the frames a little easier. The itinerary of hive-making parts is never-ending. I must write a list next time.

Now, in theory, despite a tiny bit of equipment and a quick repaint in some areas, hive one is ready and I just have to wait for the Beehaus to arrive – said to be the end of the month. Exciting times.

I was thinking today about the bees I will be getting and looked at the map to see how far away they were. It looks like a 15-mile journey to collect them and then, apparently, I pick them up in a little wooden box and drive them back. The thought of driving with bees frightens me; what on earth happens if they escape?

I would think usual, flat roads would be OK and wouldn't disturb the bees too much, but the last 800 yards will be a nightmare. Our farm track has more holes and divots than a golf course. I can only think that I will have to get out and walk up the road with them. The alternative of 5,000 bees all bouncing around before release isn't particularly palatable, especially as I will be in the firing line!

As everything is becoming more of a reality, I find I am already being a little reflective on the year so far. When I first started I felt almost embarrassed by the fact that I was going to be a beekeeper. As the year has gone on I am becoming more proud of the fact each day. Back at the start of the year I was even questioning myself on exactly what I was doing. I didn't expect everyone to be quite so interested, asking how I am getting on and seeming genuinely intrigued by it all.

Yesterday, Jo and I went over to Paula's (our next-door neighbour's) house for a quick drink to welcome Jo and Nicky. There are only four houses where we live, so it's nice to meet up every so often. We got talking about gardening, as I am often up early tending the garden before work which most people find slightly odd, and then the natural progression was to mention that I was becoming a beekeeper and that there would be a couple of hives up the lane.

They were fascinated by it and they all volunteered to 'look after them' when we go away – not sure it is the same as looking after the cat but still a very nice gesture. We spent the next half an hour discussing bees and I started to churn out all the information that I had learned; this only intrigued them more. At the end of the evening, a honeybee must have heard she was being talked about and flew in the open window. What a lovely coincidence.

This sort of thing has been happening at weddings and dinners we have been to, or just down the pub talking to mates of mine. Everyone seems genuinely interested and keen to learn more (much more so than when I start to talk about sweet peas or dahlias!). I have therefore decided to buy a second bee suit so that when people come over to visit, I can take them up to the hive during my inspections. What a lovely experience to share.

MAY 5

I have just come back from another war story session at the local pub with the beekeepers, having done another practical session at the apiary. This time they were more concerned with discussing politics and the upcoming general election than beekeepers being stung in the eye. My theory about last week being an initiation ceremony still stands. I can only presume we have passed the test and have been accepted.

I wonder what the initiation ceremonies would be like if universities had beekeeping clubs? Would they consist of being stung in the eye, going into an apiary naked covered in honey or maybe just having to recreate the bee beard?

Anyway, it really was a fascinating session this week and I probably learned more than ever before. Of the eight or nine hives that the association uses at the apiary, we have undoubtedly the 'most rubbish' hive – which is how it has been described on several occasions. Not only is it a very small and weak colony, but it's also in the oldest and most dilapidated-looking hive on a very small hive stand. With most of our group similar in height to me, this rather small hive is rather tricky to deal with as well.

I won't bore you with the details, but the queen hasn't been laying for a few weeks and there is a concern that the colony won't survive. Adam came over to debate the issues we were having about bumping the queen off. After some jiggery pokery involving a frame of eggs from elsewhere, we decided to leave it to the bees to decide what to do. A frame of eggs may incentivise the remaining bees to raise a new queen and so we left them to it, hoping for the best.

It is quite funny, when I am in the apiary even if the wind blows on my fingertips I feel like jumping 10 feet high, fearing I have been stung. I visualise the day when I get stung holding a frame of bees. Knowing me and my rather overstated reactions, my arm will suddenly fly up and the frame will crash to the ground. I think the next thing people would see is me running towards the nearest pond and jumping in followed by hundreds of bees flying after me.

It was good to finish the evening with a nice pint down the local pub, a head filled with knowledge and enthusiasm still flooding out.

MAY 10

I am beginning to worry. Did I actually order any bees or was I imagining it? Time is creeping by and each and every week, during my practical sessions at the local association, I see colonies getting stronger (with the exception of our one). I hear of swarms being caught and given to other beginners. I also hear of others who have picked up their nucs and successfully moved them to their own hives and I am feeling a little left out. Here I am, my hive is built and in position and the other one is on the way, and yet I still have one major component missing: bees.

Imagine the embarrassment of hives with no bees. I couldn't make my one pot of honey without them. Hopefully I will hear something soon.

MAY 15

I am stuck in another hotel room as the corporate world has cruelly taken me away from the family and the prospect of bees for another Saturday. I was looking forward to preparing the area for the Beehaus beehive. Surely it can only be days away now, and it's such fun thinking what could be happening very soon, but there is still no sign of the bees and concern is growing by the hour. I have emailed the beekeeper supplying them to find out what is going on.

However, all is not lost. I am here in Poland and with bees constantly on my mind, even in this rather dejected and run-down city of Lodz (pronounced 'woodge' for some strange reason), I have had some contact with bees and beekeepers.

I took my staff out for a meal last night and, as is so often the case at the moment, conversation switched to bees. I then learned that the Polish for bees is *pszczola*, or something like that, and please never ask me to pronounce it. Then I got dragged to yet another bar which I have to say, despite most people's thoughts of this former communist state complete with pot holes in the road (true), tractors driving down city streets (true) and rather bland but many government buildings (also true), was amazing. The Beer Hall was the literal translation. Something akin to a jazzy London bar and not at all what I was expecting.

I got marched to the bar where they started talking enthusiastically in Polish. I can get a grasp of most languages and I thought, given the number of Poles in London (after Warsaw, London is Poland's second largest city by population of Polish people), I might understand the odd word – not a chance. Complete gobbledygook. Regardless of what they were saying, in front of me appeared the most fabulous-looking golden beer. The guys pronounced proudly that this was 'honey beer'. Great, I thought, you can even make beer from bees! My newest hobby has now reached new levels. It has no negatives thus far and I have to say the beer was fantastic save its head, which was almost as high as the rest of the beer itself.

Fast forward to today and having been part of an exhibition which was conducted 100 per cent in Polish I was feeling a little bit like a spare part – not to mention a teeny weeny bit hung over (after the euphoria of the honey beer we had decided to try the whiskey bar down the road and eventually got asked to leave at about 4 a.m., and only then did we realise that we were the only ones left in the place – it seems my ability to drink whiskey has dramatically improved this year!). I therefore decided to go for a walk around the complex – imagine a massive World War Two aircraft hangar painted a rather dashing yellow and blue colour – and to my surprise and pleasure, I found they had a flower market outside.

It was incredible with the most amazing variety of plants on sale and all unbelievably cheap. I thoroughly enjoyed looking around and then, to my complete astonishment a clapped-out, flea-bitten camper van grabbed my attention. With its boot door open they had a small stand selling none other than Polish honey. With the rather primitive sign showing a honeybee flying around and the fact that he only had about four jars left (I would like to think that he had sold all of the rest but maybe he hadn't actually sold any) it looked so fantastic I had to go and check it out.

The two gentlemen, who looked a little bemused at the presence of this suited and booted Englishman approaching, were in their fifties, had greying hair and the most fantastic black-as-night moustaches you could ever hope to see. Honestly, it was as if they had been growing their moustaches since they were twenty, as there was just a mass of hair wedged between their upper lip and nose. I would bet they haven't been kissed for decades as a lady wouldn't even get close to their lips.

Given the fact that I had had absolutely no luck trying to speak to Polish people so far, I felt I should give it another go here and try to strike up a conversation about bees with these guys. This was after all the country where 'modern' beekeeping began with Johann Dzierzon. The blueprint of his hive design back in 1838 formed the basis of many of the hive designs used today. I was hoping that our mutual love of bees might see me through the lack of mutual language if they didn't speak English. As I approached, waving perhaps overenthusiastically and said hello, my fears were pretty much confirmed. They looked at me blankly and their lips and moustaches hardly flinched as they said *'czesc'*, or hello, back to me. My 'How are you? I am a beekeeper too' was perhaps a step too far. More blank stares as the two of them just stood there.

My attempts at sign language were also a bit lame and I felt beaten back by the moustaches. For some reason they didn't understand me when I pointed at myself and then ran around for a bit buzzing away and then pointed at the honey. I think it must have been the suit as other people were now stopping to watch. The look they were all giving me was not dissimilar to the look that Sebastian often gives me when I run around him buzzing. Maybe nineteen-month-old children and Polish gentlemen with monster moustaches have more in common than you might think.

I stopped, resigned to the fact it was never going to work. Having spent all my zlotys I couldn't even buy a jar of honey, which looked a fabulously dark colour. I left defeated but as I bid them farewell, one lifted an arm as if to wave while the other uttered *'pozegnanie'*, which means goodbye. I felt I had finally connected but it may just be that they were glad to see the back of this rather crazy Englishman. As I walked away, I did turn back for one last look at these moustached Polish beekeepers and they were just looking at each other with a rather strange expression; one then shrugged whilst the other started laughing. I must have made some sort of impression. I am just not sure whether it was a good one.

MAY 16

The experience of dealing with Polish people and their impressive moustaches is over and I am now back in the UK. It looked like the weather would hold and so Jo and I decided to pop over to RHS Wisley again as it's only about half an hour from us. We were aiming to go to areas of the garden that we hadn't been to before, including a fantastic tree that is flowering at the moment, which is quite a rare event. It is called a 'handkerchief tree' because its flowers are like handkerchiefs. When the wind picks up it is beautiful and elegant to look at and it gives off a faint and delicate aroma that is simply magical. Visiting Wisley is quite a strange thing for me. I leave it feeling inspired, jealous and a little bit humbled. What they do there is simply incredible and my hat comes off to the gardeners there – their attention to detail is second to none.

As we were walking through the allotment area, something I am always keen on doing, we came to the lettuces which, though so simple, were out of this world. They alternated green and red varieties and as I was looking at them, drooling somewhat, Jo said under her breath as if in disbelief, 'They are all identical in size and the straightest lines I have ever seen.' It was an incredible display and crikey, these were just lettuces.

Other particular favourites of mine at this time of year are rhododendrons and azaleas. Sebastian evidently loved them as well and it was only then that I realised he had the eyesight of a fighter jet pilot. Every five seconds he would shout out 'bee' and then get all excited. There in amongst the exquisite flowers were bees, darting around from flower to flower making my son's head swing wildly from side to side as if he was at Wimbledon watching the tennis. Jo and I were just smiling and watching the simple pleasure he was getting from watching these small insects buzzing around. The words 'like father, like son' were uttered and I couldn't agree more. It was really lovely to watch, but even lovelier to see him excited by them and also knowing what they were. OK, the fact that later he thought a woodlouse was a bee as well was neither here nor there...

We all went and had a great lunch and then finished the visit off in the library. The people there are always helpful, and today was no exception. As I returned one of my borrowed books about bees, the lady behind the counter saw Sebastian and recommended to us the bee books in the children's section. Things so often missing these days: a mixture of good service and knowing what people want. Sure enough, in the collection of books about insects, bugs and all things creepy crawly were some about bees. Sebastian immediately shouted 'Bee!' at the top of his voice, much to the amusement of others in the library, and then proceeded to read the book from cover to cover.

We came away with two more books on loan including Sebastian's bee book, and all very happy and fulfilled – what a great day out as a family.

MAY 23

Another week and still no bees and no word back yet. I will try to call the supplier of the nucleus this week I think to see just what is going on. I have a primary concern of actually getting them but a secondary concern that if I am able to magic them up from somewhere, they will have to quickly make enough honey for me to actually get a jar.

I am starting to put the word out that I would like to get a swarm. I suppose it is covering all bases, and should my second hive arrive I'll want another lot of bees. There seem to be a few people who are offering to help me out and given the weather of the last couple of days, by far the best of the year so far, this week could be a busy week because bees will swarm in good sunny weather. I will contact them directly this week.

The irony of all this is that I am actually really nervous about having my own bees. Any time now I could be getting a phone call that my bees are ready to go. I am not sure that I am quite ready for that level of responsibility. It is a funny sensation being excited on the one hand, frustrated at not having them on the other and then exceedingly nervous at the same time!

All in all, this is a pivotal week and I predict by the end of it I may actually have become a beekeeper and not one that is just pretending. My days of being a theoretical beekeeper will have ended.

MAY 24

The Chelsea Flower Show started today without me as I didn't order tickets in time this year. I have decided that scaling the fence is not an option so I will just have to watch it on TV. I feel slightly upset as I have gone for the last few years and loved it. However, it is all hotting up regarding the bee situation.

I heard a rumour yesterday that one of the local beekeepers was running out of boxes to collect swarms. As I've mentioned before, to collect a swarm, you basically give your association a brood box full of frames and foundation, and when they are called to a swarm they take your box with them. I am definitely going to hand over my brood box on Wednesday when I go to the practical evening. The temperature touched 28 degrees Celsius today so there are plenty of swarms around at the moment and I should almost be guaranteed some bees.

I also texted the guy who said he would be supplying my nucleus to see what he's up to, having not heard back by email.

Considering I was meant to be getting the bees at the start of May it does seem a little bit late and I haven't heard a thing. I got a phone call back almost immediately but he rang my home number and left a message. In a fit of confusion – we do have the most complicated and stupid phone set-up in the world – the message was mistakenly deleted before I got a chance to listen to it but Jo is certain it said something along the lines of, 'Your bees are OK and it will be another week.' Rumour has it that he is having a problem with the queens – God knows what problems he is having but I suppose queens will be queens.

As I said, this could be a very interesting week.

MAY 26

Today's practical session with my local association was the best session yet but there were two very significant incidents.

It all started with me trudging through the car park with my brood box, filled with frames, ready to hand it to the first person who said, 'Ah perfect, just what I need, I have a swarm for you.' However, I doubted it was going to be that easy, and it didn't help that I was finding the simple task of walking with a brood box difficult enough. After taking only two steps, I tripped and fell into the bonnet of my car promptly putting a great big scratch along it. What a great start.

One of the ladies in the group, Suzy, was there to greet me. She has actually been reading my blog and asking about my lack of bees, and she said that I should speak to Richard at tonight's session. This is the same chap who was running out of boxes and my mentor, Adam, had also recommended I see him tonight. Great news, I thought, only to find that Richard wasn't in fact there. Just my luck.

In fact, it got even more farcical when I found out that almost all of my fellow beekeepers had now got their bees. Bearing in mind I had thought long and hard about where I would get my bees from and didn't follow the crowd, and opted for a guarantee of bees by buying a nucleus, I was spitting feathers as it slowly dawned on me that I had got it all wrong. This often happens in supermarkets when I pick the wrong queue, having analysed them all for some time and thinking I had picked the one that seems to be moving the quickest. It's just typical that this year is a very swarmy year, too many swarms around, while my nucleus maker is having a few problems...

Apparently Eddie, another beekeeper, would be happy to take on my boxes to catch a swarm and so I went to meet him in Richard's absence. Nice guy, but he took one look at my box and said nope, I had the wrong type of floor for a swarm. I was slightly dumbfounded and saw others around me smirking. According to Eddie, swarms prefer a solid floor and not an open-mesh floor, which I had been told to buy from pretty much every other beekeeper. Stumped once more, here I was with what I believed to be a five-star hotel of a bee box and it wasn't quite right, for Eddie at least.

It was evident from others later that Eddie is a little rigid in his beliefs about swarms only settling in certain boxes and won't hear of any other opinion. I couldn't really argue with him.

To cut a long story short, once we were at the pub and I was being consoled by the others, Adam said he would take my box and deliver it to Richard but did mention that Eddie wasn't wrong about the type of floor to use. He stated that it is generally accepted that solid floors are better for bees when housing a swarm than an open-mesh one. Usually people will swap a few weeks after housing the swarm. Result. Either way, maybe I will get my bees this week after all.

The other story started with Tom, my worldly wise tutor, suggesting we all learn how to pick up the bees. In went Tom with finger and thumb and just effortlessly picked one up.

This was a step too far for me. Picking up bits of wooden frame with bees on them was one thing. Actually picking up a live bee complete with a weapon of mass destruction was quite another. Tom suggested starting with drones as they have no sting and they are a little fatter and therefore easier to pick up. I was still not convinced, but in launched Richard and other fellow group member, Andrew, without even thinking. They both harpooned a bee immediately and looked at them studiously. Neil, a nice guy and the only other in the group tonight not to attempt to pick up a bee, looked at me and I at him; gladly I could see that he was as sceptical as I was.

I couldn't be made to look like a wimp and so I made several willing advances towards the bees, finger and thumb extended and doing the occasional pincer action to show intent. Each time I got to within a whisker of a bee it moved. Damn moving things! In fact, I was breathing a deep sigh of relief each time they moved but kept up the pretence by making huffing noises and claiming the drones I was going for were obviously a fast strain. I'm not sure I fooled anybody.

I am determined, next week I must pick up a bee, even if it does turn around, shout out that I am trapping its leg and then proceed to sting me. I need to get it out of the way sometime, I suppose.

All in all it was a great session and one really worth going to.

MAY 28

People have many traits that they take from their parents. I am fortunate to have picked up the best sides of both parents and am pretty laid back, quite decisive and a relatively good communicator (or so I am told). However, one of my many faults is impatience. Now I am about to pay for it.

At the start of the week I felt it would be a busy week, and I wasn't wrong. I hadn't heard anything about the nucleus and the rumour mill was suggesting the first week of June, contrary to what the guy had told me. So, having become excited and given my box away to catch a swarm on Wednesday, which was apparently imminent, I received a phone call today to say that my nucleus was ready to be picked up. How typical and how completely ironic. This leaves me in a position of having some bees but now I have no hive, a complete turnaround on the last few months.

In my head I had calculated that I could give away my brood box to catch a swarm and in the meantime receive delivery of the Beehaus. I would then transfer the bees from the nucleus to the Beehaus instead – perfect, I thought. In fact, having talked to Omlet today I now realise this is a pipe dream. It is not due to arrive at Omlet until the end of next week as there have been some last-minute modifications, which they want to update before giving me the hive. My shoulders sagged and I heard my mother's voice saying 'told you so' reverberating around in my head. I have therefore had to pass on the nucleus offered so he can give that one to someone else. I will just have to wait another couple of weeks to get the next round. Hopefully by that time I will have a swarm safely ensconced in the National hive.

Trying to get a guarantee of bees I have put myself again in a position where nothing is guaranteed! Oh well, let's soldier on and hope that an army of bees is swarming somewhere around Reigate, that a nucleus of bees in Farnham is brewing nicely and that Omlet is confident that the Beehaus is on its way.

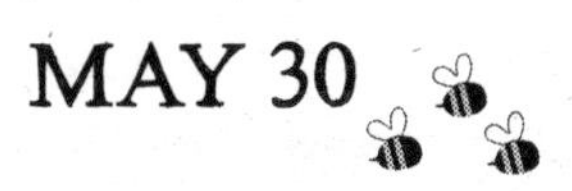

MAY 30

Well, today has been a most interesting day.

Sebastian met up with all his mates today and was having a whale of a time just larking about. It is amazing the transition in just twenty months, now seeing the little characters coming out and long-term friendships settling in. At one point, Sebastian walked up to Jo, bottom lip out and quivering. 'Bee' was the word being uttered as his little eyes started to well up a little. Immediately we started to see what the matter was and decided he wasn't just uttering letters of the alphabet at us. We started stripping him off and checking he was OK and sure enough as we removed his arm from a sleeve, a little bumblebee just flew out, completely unharmed. Sebastian very quickly started smiling again, shouting 'Bee!' at the top of his voice and excitedly jumping up and down. After a check, no stings were discovered and we all concluded that the bee had been playing hide and seek for a little while.

We got back home and put Sebastian to sleep and I then asked Jo to help me film my first lighting of the smoker in readiness to put the clip on YouTube; a very piquant moment. I realised I hadn't lit my new smoker yet and felt I had better just test it out. Thankfully, it went very smoothly and lit first time. The smell is something indescribable but already cemented in my memory bank. Jo was impressed, I think, and I felt caveman-like, much like most men do when they light a bonfire.

Our lovely neighbours Jo and Nicky saw me light the smoker and were intrigued and so I popped around and showed them what it was all about. Walking back from their house I received the phone call which was to change my life. It was about 7.30 p.m. by this point and a lovely evening with a few clouds in the sky and a number I didn't recognise popped up on my mobile. As soon as I heard the voice and the name Richard I knew what it was about. Richard – the 'swarm-catcher' for the Reigate area.

This could mean only one thing. Having given my box away on Wednesday to Adam, not only must he have given my box to Richard as he said he would, but he must have a swarm for me. Here was 'the call'.

Apparently some bees had swarmed into a nursing home's garden. Not only that, but they looked like they were Suzy from the beekeeping association's bees, who were swarming for the second time in seventy-two hours. On the Wednesday she mentioned that she had some rather prolific bees swarming, and here they were, ready and waiting for me. It sounds as if they were simply running out of room in her hive and had to get out.

Here I was talking to an experienced beekeeper who was having to deal with all of this late on a Sunday evening and it was quite a surreal, one-sided conversation. It went a little like this:

'I have a swarm for you, I have your box, they are fanning at the entrance and it looks like it was a successful catch. Can you come and collect them?'

With Suzy shouting various bits of information in the background as to how to get there it was all over in about a minute. At least he proved that you don't necessarily need a solid floor for a swarm – that was my first thought. I was suddenly incredibly excited but also a nervous wreck. Eight months' worth of reading, writing, investigating and learning was about to pay off with this simple and unemotional one-minute phone call. It all seemed so easy and straightforward with one exception: I still didn't feel remotely ready.

While I was talking I realised there were several things that I still hadn't done. Firstly I needed to move the hive position as it was too near the shed. Secondly I had no sugar to make up a sugar solution to give the bees upon arrival to encourage them to 'draw' out the comb. In their new home will be the frames with wax foundation strips and you need them to build the comb onto these strips in which the queen can start laying. These also provide a base for the bees to store pollen or nectar so if they don't get this job done first there could be a few problems settling in. It is my understanding that this is standard procedure for newly swarmed bees but perhaps most importantly, I didn't actually have a feeder to give the sugar solution to them. I would have to ask Suzy if she had one and whether I could borrow it. These are usually large plastic or wooden containers that sit directly on top of the hive and will hold the sugar solution. There are a few different sorts of feeder but the bees will have easy access to the solution and make use of it in the hive. Within a few weeks, having topped it up several times, you should apparently be able to take it off because the bees will have established themselves in the hive.

I finally left home at 7.50 p.m. with my heart beating a little faster than normal. About five minutes down the road I realised I hadn't brought the address with me nor brought any gloves. Typical. It didn't matter too much as the address was cemented into my mind and I was not going to be touching the bees today, just the hive.

I got to the nursing home just after 8.30 p.m. and dusk was settling in. I drove in and Suzy was there with her daughter Laura, all dressed up and ready to go. We exchanged pleasantries though I felt a little awkward, especially as I was about to steal their bees! While we waited for Richard the swarm-catcher, I was feeling strangely grown up and couldn't really put my finger on exactly why.

Richard's car pulled in and out stepped a gentleman in his early sixties. I am not sure what I expected a swarm-catcher to be like but I am not sure I expected him to be so charming and normal. From our discussions I was expecting him to be quite precise and almost military-like but this was simply not the case. Tall and gangly like me, with an air of calm and quiet authority, Richard went about explaining what had happened and what we were to do next.

Walking through the grounds of this care home in my beekeeping outfit, in the distance I could just about make out the hive situated under a 5-foot-tall tree. As Richard had pointed out, these bees could not have been any more accommodating. Surrounding this small and rather sad-looking tree were some 50-foot monsters, which would have needed extreme climbing gear and probably some of Reigate's finest scaffolders to build complicated platforms to get to the swarm. However, they had very kindly picked this small tree.

So, on the ground was my nice little hive, on top of a white sheet. It was immediately apparent that my open-mesh floor had been replaced with a red, rather old-looking solid floor which had been screwed in place using a metal plate. Richard mentioned that

Adam had kindly done this as he felt it had a better chance of keeping a swarm in my box, which I thought was very kind of him. It just meant that I would have to change it back again in a few weeks when they were settled in.

Earlier on in the day, Richard had cut off the branch holding the swarm and held it over the box and given the branch a tap. The aim was to knock as many bees as possible into the hive along with the queen. The branch was then laid in front of the hive and on top of the sheet, and the stragglers could walk into the hive, sensing the queen was already in there. We simply put some wire mesh over all the openings and held it in place with drawing pins, before carrying the hive to the car; all the while I could hear the buzzing of the bees, which were probably wondering just what on earth was going on.

I have to say, despite knowing that the bees couldn't escape, it is one of the weirdest sensations putting a beehive in the boot of my car. I cannot think of one more stupid act in the world. To top it all off Richard, who at this point I deemed wiser than one of the Wise Men, gave me a long diatribe about what to do next. One statement slightly perturbed me. Just as I was leaving he said, 'Drive slowly, don't go round too many bends and avoid bumpy roads.'

Here I was, driving a car complete with a full beehive and I have to navigate some of the windiest roads in Surrey to get back home. To finish, I then have to navigate a pothole-ridden farm track to get to my house. To make matters worse, it was now getting dark and I knew that I had at least thirty minutes of driving ahead and so I knew I would be putting them in their new home in the pitch dark. I wasn't feeling particularly confident.

I arrived home unscathed and surprisingly even the farm track seemed fine; they didn't even buzz. I stopped the car outside the house and ran in with the newly borrowed feeder that Suzy had very kindly lent me. Having realised only two hours earlier that I was wholly unprepared, I asked Jo whether she could find any sugar in the house. Fortunately we had some brown caster sugar, not great as I think they prefer white, but would do the trick for tonight at least. Therefore I started to make up the sugar solution by mixing equal parts of sugar and warm water. Very soon it was ready.

As quick as I could I then made my way back to the car complete with smoker and gloves and drove up to the allotment. By the time I had got there it was absolutely pitch black and so I decided that this installation would have to be by car headlight.

I have seen plenty of YouTube videos of people installing their nucleus of bees to prepare me for this moment. However, most of these videos show the charming sight of people in nice soft sunlight putting the nucleus into position. Here I was living the moment complete with torchlight and car headlights. I felt somewhat cheated.

I eventually got the hive onto its stand though, and started the procedure of setting them free. I wasn't too concerned as it was dark and they would probably want to stay where they were for now, but I got suited up just in case.

I started with the mesh on the crown board (the very top of the hive), as this was where I was going to place the feeder. I soon realised I couldn't actually remove the drawing pins holding the mesh in place. Typical. I had to go back in the shed and get my hive tool, which, when I got there, despite its most psychedelic colouring, I couldn't find for love nor money and instead had to settle for a knife.

I finally got the mesh off and put a temporary cover over just in case but then realised I couldn't undo the strap holding the hive together – this was put in place to stop the hive moving apart during the journey. DIY was never my thing and so to see a ratchet-type contraction holding it all together was rather perturbing and I started kicking myself having not asked Richard about this earlier. Truly, this must have been the longest part of the moving-in process. It took me ages to work out what to do and even then I only managed to loosen it a little and pull it gently off the hive. This should have been a thirty-second job but it must have taken fifteen agonising and frustrating minutes. Anyway, it enabled me to get the bright green feeder onto the hive and I let out a small sigh of relief knowing that the bees would be fed at least.

Then all that was left was the small matter of undoing the front mesh, which was, to be honest, the nicest bit of the process as it was relatively easy. As soon as the mesh revealed the small entrance, a little bee popped his nose out and under torchlight had a little look around. Shortly afterwards another popped out to see what was going on too. I decided to leave them to it.

I drove back to the house at about 10.30 feeling quite satisfied that they were all right. I got in, filled Jo in on all the details and walked upstairs to the study, which is where I am now. I have some funny feelings going through my head, not least the fact that I can now say to people that I have bees. I can join in with the discussions on my Wednesday night sessions about how lovely they are and can swap stories.

A thought that has also just occurred and it is something that Laura, Suzy's daughter, told me before I left the care home. 'Now,' she said, 'the queen is called Nefertiti and her bees are collectively called Patricia.' Therefore my queen, if I am able to find her, will be called Nefertiti. I did ask whether the boys had a name and she said no. Therefore to stamp my own mark on this swarm, I will call my boys Paul. Why? I do not know.

I have heard that some beekeepers like to talk to their bees but never name them; in fact I have heard some say that talking to bees is a tradition that should be upheld. Beekeepers should apparently tell the bees all of the family news so they know what is going on. Should someone in the family die you should go and tap on the side of the hive and tell them what has happened to stop them swarming. Hopefully I will not have to be doing that anytime soon.

Suffice to say I was right all along, this has been an eventful week and with only one hour remaining of the week and Monday beginning once more, I can look people in the eye and say that I am a beekeeper. That's a nice feeling.

MAY 31

The day after the night before... I slept awfully and I have to admit it was all self-inflicted. While I was writing the diary I had a couple of gin and tonics to soothe my overexcited brain. What I had just experienced was bonkers and my mind was working overtime, and despite the alcoholic intervention, I struggled to sleep.

I woke up knowing that a few minutes' walk away there was a hive full of bees waiting to see me this morning and so I wanted to get up there as soon as possible just to make sure that all was well. After all, last night everything was done by torch and car headlight. It probably meant that I had put everything on wrong, or left the roof off or done something else equally stupid.

Fortunately when I got up there all seemed well. A few bees were just showing their heads outside the hive wondering where they were. Over time I noticed some of them actually plucked up the courage to take a little flight. It seemed they would fly about a metre away and would then get a little bit nervous and fly right back again. I put it down as the equivalent of watching your little ones learning how to walk. They stand up, hold on to the sofa or an equivalent, take a few tentative steps holding on, let go for a step and then immediately grab for the sofa once more.

I understand now that these are called orientation flights and they are just getting used to their new location. Gradually these flights will get further and further away from the hive until they are satisfied they know exactly where they are going. Apparently they use markers in the landscape to guide them to and from the hive; amazing.

I lifted off the roof to take a look at the feeder which, to be honest, I dealt with a little bit hastily last night. It all seemed fine but I had no idea what to expect. I must find out just how quickly they should take on the syrup so I can gauge if what they are doing is normal. The only worrying thing was that I saw two dead bees floating in the syrup. Having just done some research though, I found out that this is pretty normal and I shouldn't be too worried. What a way to go, swimming in sugar – it must be our equivalent of drowning in a pool of money.

My concerns abated, Jo, Sebastian and I went about our daily duties and today this involved meeting up with a group of my old school friends for a barbecue. We packed up the car complete with Sebastian's bee backpack which Jo had recently bought him and got going. It looks amazing on him and he loves running around pretending to be a bee.

Ten minutes into the journey there was a little buzz and a honeybee landed right next to Jo's hand. Whoops, one got left behind last night. I lost a few brownie points there! Jo let down the window but I doubt he would find the hive now. Sad, really.

JUNE 1

I had a minor heart attack last night. Just before I got into bed I checked my BlackBerry and there waiting for me was an email from Adam which I opened and my heart sank. It simply read:

'Any white sugar is fine, brown sugar gives them dysentery.'

Oh my God, what have I done? Dysentery, or serious diarrhoea as it is probably better known, is colony threatening so I had made a major mistake. I had fed them brown sugar as I didn't have anything else in the house. I checked the label of the packets that I had used and found out that it was actually 'light brown muscovado' sugar. This meant that at some ungodly hour of the morning I was beginning to fret about whether light brown sugar would be worse than dark brown sugar. I can tell you I have never before sat down and considered this argument but the difference between these two sugars kept me from sleeping particularly well.

I woke up this morning and came to the conclusion that I had to reply to Adam first before I did anything and so I left the feeder on and emailed Adam for confirmation. A little later, just as Jo and I got into the car to go to my mum and dad's he emailed back and basically stated that light brown was just as bad.

So, on the way to my parents' house, Jo jumped out at the supermarket and got some more sugar for me. In she went to Sainsbury's and came out with two rather heavy-looking bags and I instantly felt a lot better knowing that I could now get it all sorted.

Sadly, as I write this it is too late as we arrived home a lot later than we thought – too late to go out there and open it all up to remove the feeder. I am sitting here worrying that these poor bees have not got the right feed, but I plan to wake up nice and early and sort it all out. Here's hoping that they do all right tonight and I haven't done any lasting damage.

I started my beekeeping career knowing I am someone who learns from my mistakes and these last two days have lived up to that theory. A lack of preparation has meant I have seriously messed up again and I am feeling very guilty and slightly annoyed at myself. Usually I am the person that suffers and not a hive full of bees. It is not fair and is unacceptable on my part.

JUNE 2

I got up early this morning to sort out the mess from yesterday and walked into the kitchen with the bags from Sainsbury's. They were full of sugar, thankfully, but when I pulled it out of the bag I realised that Jo had bought unrefined golden caster sugar, which was not exactly white and so I am again concerned. I take full responsibility here as I wasn't particularly specific and just said, do not get brown sugar. I never realised just how many different types of sugar there were. Although I know that white, granulated sugar is best I have decided to take a chance as it was only a little off-white and so I think it will all be OK.

As I did a few nights ago, I melted the sugar into warm water and, after letting this cool, transferred it up to the hive. I had a quick peek outside the hive and they all looked like they were flying around again, which was good news. I hadn't killed them all then. I went into the Man Shed and lit the smoker and then headed out to the hive to change the syrup.

I felt a little bit better after doing that but I still had a nagging suspicion that something wasn't right. I felt it was an apt time to call upon the oracle once more and popped Adam a quick email. Sure enough, about half an hour later my suspicions were confirmed on my way into work – golden sugar was just as bad. For Christ's sake!

Therefore my day involved getting some white sugar and then rushing back from work to deal with the same issue once more. It didn't help that tonight Jo and I were going out for dinner, having organised a babysitter for the evening. Before I could do anything I had to rather subtly suggest that I had to melt down some sugar and feed it to the bees... again. I cannot say it was a passion-killer but I will say that I got a rather 'I cannot believe you are doing this' type look.

Anyway, with the feed safely in place we rushed off to dinner with me feeling a little bit relieved, though still quite guilty, that the situation had now been resolved in the short term; I only hoped it would have no lasting effect on the bees, though I'm not actually sure what to look out for. I certainly shan't be thinking about sugar in the same way ever again.

JUNE 3

Crazy, isn't it, how time flies? I cannot believe it is the third of June already. Today I got up early to check on the bees, to make sure they were alive and, ultimately, that they were still there.

Apparently a swarm of bees is still likely to swarm again if they don't find the new location suitable. Fortunately they were still there and a quick look at the hive entrance showed them to be more active than normal, probably due to the fantastic morning we were having already. The sunrise was beautiful.

Anyway the astonishing news today was that I may not actually have Nefertiti, the queen I was told to look after by Suzy's daughter, Laura. I may have Cleopatra, who was the mated queen she received in the nucleus she had purchased six weeks previous before they swarmed. I am pleased about the possibility of a more memorable and easy to pronounce name! This was due to a bit of guesswork on the part of Richard, the swarm-catcher, who, when inspecting the bees she had left, discovered a virgin queen in her colony.

This therefore could be interesting as my colony could get going a lot faster – she was a prolific layer, hence why Suzy's colony kept swarming. That would be nice as I originally feared that Nefertiti would have to go out and get mated. So fingers crossed as more bees means more workers, which should mean an increased chance of a jar of honey this year.

Although I've left it rather late in the season, if Cleopatra produces as rumoured, I could still be in with a chance.

JUNE 4

I went up to the beehive today and had a quick look outside. There they were, flying in and out happily – amazing really, and slightly mesmerising. They seem so content to just keep on going in and out, in and out, it is quite therapeutic to watch. The great news is that I saw several bees flying into the hive with huge great pollen sacs attached to their back legs. In short it means the hive is content and is going about its duties. Great news. It also means that I haven't killed them off with the brown sugar. Happy days.

JUNE 5

The bees seem pretty hungry, I have to say. In only twenty-four hours they have been through 1.5 litres of sugar syrup. That seems a lot to me but this is quite normal and still a good sign, as is flying with pollen sacs, and apparently I just have to keep feeding them. Therefore I gave them another 2 litres this morning. It will be interesting to see how quickly they get through that.

One event today made me realise I have a long way to go. I went up to the allotment for a couple of minutes to see how they were getting on; it must have been about midday and so I knew they would be a little bit busier than during my morning visits.

There I was standing there in my shorts and T-shirt – it was a beautiful day, the warmest this year so far – and I was just watching them fly in and out. There was a lot of activity and yet again there were a few flying in with pollen sacs, which again was reassuring.

Every time I have approached the hive I have wondered just how close you should get before you should consider wearing a bee suit. I have come to the conclusion that if you are just watching them you should be OK, and I have also taken the decision not to wear a suit when adding more sugar syrup as there is no direct contact with the bees.

Today I had a small warning and I think I need to be a little more careful. I was there watching them, about three feet from the hive entrance. It was relatively peaceful until one particular bee decided to fly at me. It is quite disconcerting when this happens, purely because of the sound. Usually it is said that bees buzz at a tone equivalent to a middle C and you can tell when they are a little bit touchy as the tone changes. This little bee stopped about eighteen inches from my nose and just hovered there for a bit pitching at slightly over middle C as if sussing me out. I still wasn't too concerned but this all changed when it decided I was a threat and flew directly at me.

It was similar to those cartoons when a wasp rears up before then flying straight at the victim. Its speed caught me off-guard a little and made me stumble backwards over the loose ground of my allotment. For some reason I also instinctively swiped at the bee with my right hand while trying to stay balanced with the left. I wasn't very successful and this small bee, probably no more than 5 millimetres in length, floored a rather tall human being.

It didn't stop there and made a beeline (literally) for my head. I immediately jumped up and started running as I heard her go at me for a second and then a third time; my arms were both pumping and wildly flailing as I tried to run away from what I now know to be a guard bee. Their job is to guard the entrance of the hive, which this one was obviously doing very well. It was quite scary, though it must have been quite funny to look at from a distance.

Suffice to say, I learned my lesson but felt a little bit of a wimp having not faced up to a duel with a bee. I hardly even stood my ground. So we all know who the boss is now!

JUNE 6

I paid a visit to Farmer Ray today but took Sebastian for backup. Being the scary guy that he is – in spite of being very nice when I asked if I could keep bees on his land – I felt that he wouldn't verbally or physically abuse a guy with a child on his shoulders and may take pity on me. It's bad, isn't it, that I use my child as a human shield; but then you haven't seen the gun that Farmer Ray carries around with him.

We approached his gate, which I can just about peer over, and there he was, fast asleep in the sun. I was gutted, I had spent five days building up the courage to walk up the road to speak to him and there he was fast asleep. I needed to discuss the delicate matter of him cutting grass in the field near the hive. The last thing I would need is Farmer Ray allowing me to have a beehive in his field only for them to attack him if he got too close with his tractor. Though they wouldn't hear the tractor they would probably feel the vibrations of it going past and might come outside to investigate. I was going to suggest that I use a strimmer while wearing a bee suit and cut the grass near the hive so he didn't have to.

I didn't dare wake him and so had just started to tiptoe away when Sebastian, who was about two feet above me and could see considerably more of the garden, shouted 'Woof, woof!' at the top of his little voice. As much as I love his fascination with animals at the moment I have to say that this was not one of those endearing moments.

In a flash Ray jumped up, grabbed his gun and looked around. I quickly looked away, but not because I didn't want to be seen

– there was something altogether more worrying. Farmer Ray spends all day dressed head to toe in blue overalls and a flat cap. I have never, in four and a half years, seen him in anything different. He was now standing there, dazed, flat cap on, but evidently before he fell asleep he had undone the zip on his overalls and the resultant view was something I will never forget. His blue overalls were around his ankles and he was wearing a rather fetching string vest and the most beautifully decorated and colourful Bermuda shorts. In an embarrassed rush he dropped his gun to the floor and quickly pulled up his overalls and then strode over saying, 'Ah, just the man I wanted to see' in a slightly less scary way than ever before. His imposing aura had been quashed. This was now Farmer Ray the 'floral-shorts-wearing farmer'.

All went well and he was more than happy to see I had taken delivery of the bees, and wasn't too concerned that they would be close to him when he was cutting the grass. It was a great result on many levels. As I was about to leave, Sebastian ran through his gate and started chasing his chickens around the garden. Ray and I started to laugh and I really feel we bonded at that moment. He is really a very lovely man and as I was learning, if you are honest with him, he is more than accommodating. All in all a worthy meeting – and one I shan't forget for some time.

JUNE 7

It's pretty horrible outside today now that the weather has turned after three days of absolutely stunning sunshine. In a way, though, it is quite nice as it means that the garden gets watered for you and also gives me a little bit of time to ponder things. With Sebastian put to bed I have a chance to reflect on what it is I have done this last week. There is one problem, however; I have no real memory of what it is exactly I have done with the bees.

Now this isn't as daft as it sounds because everything has happened so quickly. I know I went to get the bees exactly eight days ago and that I messed up pretty badly giving them the wrong sugar but aside from that I cannot remember exact details. This is why beekeepers always need to fill in what are called 'hive cards'. For people like me who have a memory like a sieve and cannot retain important information for more than an hour, these little cards are imperative.

I first saw a hive card on the course in the winter and within the hour had forgotten about them. I then saw them again during my practical sessions when I realised that they were actually quite important documents if you wanted to have any record of what was going on. It was dawning on me how important it was to document aspects of an inspection like whether you had seen the queen, how many frames the bees were on, whether you saw any eggs, was there any sign of disease; the list was endless and hence the need for a record. I had, of course, completely ignored their importance and just got on with it. Considering I had received the call last Sunday and then within the hour I had my bees, I was quite underprepared.

Today I downloaded a hive card and printed a few out and started to use them immediately. You can also get online hive cards that you fill in using your phone, which sounds pretty clever but to be honest, I am not sure of their practical application. I am also not sure what the bees would make of a beekeeper tapping away furiously into his phone in one hand while holding a frame of bees in the other. Can you imagine what would happen if any honey got on the keypad; it would take the term 'sticky keys' to a new level and rather than recording '1' queen I could end up with '11111111111' instead.

To my knowledge, despite my cock-ups with 'light brown' and 'golden' caster sugar at the start of the week, I think they have gone through at least 5.5 litres of sugar solution, which sounds quite a lot. However, I suppose, if you break it down it may not be so much. Let's assume I have approximately 5,000 bees now in the hive; that means they would each have taken about a millimetre of sugar solution each... Doesn't sound as much then, does it?

JUNE 9

What an interesting evening. Tonight I attended the 'Big Buzz' event hosted by the organic people at Abel and Cole, deemed an 'eccentric bunch' by the founder Keith Abel. Eccentric they may be but I have to say they pulled it off exceptionally well.

I approached the venue, in the shadow of the beautiful Battersea Power Station, and was met immediately with a choice of Prosecco, white wine or rosé. As I walked through with a nice chilled glass of Prosecco, I was struck by several factors. Firstly, the average age of the people in the room was easily twenty years younger than I was used to at beekeeping-related events, none of them looked remotely like a morris dancer and there was even a jazz band in the room. I have a feeling my association comrades would have fainted by now at the shock of it all as this was hardly the demographic of beekeeper or style of our local meetings and events. It just shows the difference between rural and urban beekeeping trends as well as the scope of interest that the hobby is receiving at the moment.

It felt slightly weird arriving on my own especially as the others around me seemed a pretty cool bunch – something I had not been expecting – but I was fortunate to bump into a Twitter friend of mine @helpsavebees, otherwise known as Damien, and his guru in the design world, Liz, otherwise known as @mizzlizzwhizz. Damien, in his free time, runs a charity to 'Help Save Bees' and is passionate about everything to do with their plight; he is concerned with all bees from honeybees to solitary bees and, a particular favourite of his, the bumblebee. He therefore raises money to help other charities like the Bumblebee Conservation Trust and generally raises the profile of the plight of the bees to others like myself.

Being passionate about the topic he really opened my eyes about the situation they all face with regard to the diseases, which are so often discussed at the moment. He also taught me a lot about the problems we are facing with pesticides and neonicotinoids in particular causing great problems with bee populations. There is a thought that neonicotinoids, which are chemical nerve agents used on our crops to control insects, have an indirect effect on bee populations. You could see that Damien was especially passionate about this topic telling me that the pesticides were working their way into the bee ecosystem by infiltrating the pollen and nectar of plants and as a result being taken back to the hive. It is said that neonicotinoids interfere with bees' ability to navigate, which is obviously catastrophic. It was so different hearing it first hand from people that deal with bees and beekeeping everyday compared to reading a book or newspaper article about the problems.

Anyway, the talks got underway, introduced by the charismatic founder of Abel and Cole, Keith, who started the evening mingling with a glass of wine in one hand and a bottle in the other, filling up glasses as he went, good man. He gave a basic welcome speech and introduced Steve Benbow, the founder of The London Honey Company, a pretty successful business with 850 hives dotted around the country. Steve recounted his own experience of starting out as a beekeeper, including a very funny story about why, in his second year, putting a hive on a barge on the River Thames was not a great idea. Tidal movements caused havoc with the bees' natural GPS systems, it seemed. Imagine arriving home and discovering that your home was 10 metres below where you left it! He discussed his hives that are now placed all around country, including those situated on the roof at Fortnum & Mason, not to mention the roof of the Tate Modern. Hive inspections must be amazing from up there.

One of the highlights of the evening was the honey tasting. Being a slight 'honey-phobe', this was most interesting because I haven't tasted very much honey, but here Steve had presented four different types on a plate for us to compare and savour.

I had never really considered honey much before I started all this and certainly hadn't thought about the amazing variety available. I suppose you just get used to seeing honey on the supermarket shelves or at the back of your kitchen cupboards. Before tonight I just thought honey was, well, honey. In front of me were oilseed rape honey, lavender honey, 'urban' honey and heather honey, and the difference in consistency, colour and smell was immediately apparent. Despite the fact they were all on a paper plate and I was about to taste them using a wooden lolly stick, I couldn't wait to jump in.

Having had a small taste of each of them in turn, in between explanations from Steve, the one that I was most amazed with was the oilseed rape honey. It was the colour and consistency of lard; most unusual and not entirely palatable, and not at all like the taste of honey that I remember so vividly from childhood. The urban honey was far more reminiscent of what I remember honey to be like. As a child I didn't really have a lot of honey but I will never forget the sweet smell as I opened the jar. This urban honey definitely had that distinctive smell I remember, but there was also its consistency. It was a runny honey and had a beautiful golden yellow colour that only added to its appeal when I went on to taste it. I was instantly reminded of the honey I had put on my toast as a child and with all my nostalgic memories flooding back; it couldn't fail to be the winner in my book. I may not have particularly liked honey when I was younger but the memories of putting a knife into a jar of runny honey and seeing it spill out onto the toast, and everything else around me, were good memories. That one taste of urban honey on a wooden lolly stick brought all of these fond memories back to me. I never expected honey to be quite so evocative, and this one moment made all the effort of going to London tonight well worth it.

Some were treating the event like a wine-tasting evening with discussions all around me along the lines of, 'Oh yes, I get the taste of lavender in there but George, do you get the subtle aroma of lemon as well?' I don't know if it was just because we were in London and whether it was all for show or if they really did detect these tastes but either way it was quite funny listening in.

Both the urban and lavender honey were surprisingly very runny, like water. By comparison, the heather honey was practically a granular sugary mass but had an incredible taste and texture to it, something I wouldn't have associated with honey previously. I have rarely experienced such an explosion of taste and it was divine.

Given all the different flavours and textures it left me wondering about my own honey. It is lovely to think I will have honey derived from my own garden. Given the variety of flowers I would think it would be more like an urban runny honey but I am not sure. Maybe another road trip is on the cards at some point to really see what my bees will be foraging on.

All in all, it was a very entertaining evening and all the more so for meeting up with a few friendly faces. One of them was Alison Benjamin, a columnist for *The Guardian* and co-author of *A World Without Bees*, which is a really great resource for anyone who wants to know about the problems bees are facing. I felt a little in awe when I bumped into her but she had seen my blogs and Facebook page (with my review of her book which no doubt helped) so it was a great discussion in the end with someone who is obviously all about helping the bees.

Suffice to say, well done Abel and Cole. I am going to bed now as it is far, far too late.

JUNE 13

I am a man of integrity and honesty and therefore I have to admit I am writing this piece with a glass of rosé to my right (yes, I know, not very 'beekeeping') and a leftover piece of cold pizza (with a bite taken out from a cheeky breakfast this morning) to my left. These are the leftovers from last night's World Cup match between England and the US, a rather disappointing result I have to say. Therefore I thought I would write this watching Australia beat Germany; sadly this is not proving to be the case, however, with Australia 4–0 down and with only ten men left on the pitch!

What a momentous day. You are meant to leave the bees for a couple of weeks to allow them to start building the colony with the only intervention being constant feeding. After all, they have to build a new home and get the queen laying eggs as soon as possible to keep the colony going. Therefore I have done what I am told – there is a first time for everything – and today was the day earmarked to look inside.

I wanted to film the event for posterity but decided early on that asking Jo to pop up to the hive would not be a husbandly thing to do. Richard from the beekeeping association stepped in. He agreed to help out and film this exciting first inspection of my own bees and generally hold my hand through the experience – absolutely needed, I have to say.

There are not many things that scare me, and I've been fortunate to have grown up with the confidence to give everything a go. There are two things that have never sat comfortably with me, though. Firstly horses. I am particularly scared of horses. When I was a child, one decided my mop of blond hair looked far too inviting and bit off a huge clump of it thinking it was hay.

Aside from horses, bees have always held a slight fear and here I was about to open up a hive – and on film. That made as much sense as touching a horse's rear end as I walked behind it.

Feeling very nervous, I opened the hive, following the correct procedure of smoking in all the right places. The first thing that struck me was the sound. There was this amazing buzzing sound coming from the hive. Yes, I knew, of course there would be buzzing but this was really BUZZING! It didn't just sound louder. I know that bees buzz at middle C but this sounded an octave or two higher. Bearing in mind I have been using the small, almost runt-like hive at our association, this sound was an indicator that perhaps there were more bees than I was used to and also that they were perhaps not as placid as my usual bees. As I continued the inspection it appeared to be the latter and I was feeling a little stressed as a consequence. I can't really describe the feeling of bees, almost kamikaze-like, flying straight at your veil in full attack mode. Attack after attack came as they obviously weren't happy I was there. It felt quite uncomfortable and something I hadn't really expected.

I will not describe all of the inspection as it would go on for a bit but aside from the kamikaze bees, it was all looking good. I reckon I have about seven frames of bees and some frames where the bees have been building out the comb ready for the queen to lay in or to put stores (this is usually termed as 'drawing out' comb). I am really pleased with this because I also saw sealed brood and evidence of eggs and larvae. Therefore, though I saw neither Cleopatra nor Nefertiti, I know a queen is actively laying. The mystery continues as to which queen I actually have, but I am not overly concerned as long as she is laying.

It is incredible to see experienced beekeepers look at frames at the association evenings. There I was today studying each frame for several minutes trying to find the queen whereas an experienced beekeeper like Adam would literally spend a second looking and go, 'There you go: Queenie!' I must perfect this art.

Aside from signs of a laying queen I also saw some sealed stores of pollen and honey, which is great. Looking at the newly generated comb is just magical. Understanding that I have simply put wax strips into a box and within a week or two the bees had made the most beautiful honeycomb is quite astounding really.

Having put the hive back together again and taken a deep breath, I thanked Richard for all his help and tried to reflect on exactly what had happened. It may sound strange but in a way I am quite relieved that my first inspection is over. Not only was I nervous but they were far more feisty than I had expected. Maybe they were feeding off my nerves?

The enormity of the situation is beginning to hit me. Last week I was almost a beekeeper and today I actually became a beekeeper.

JUNE 15

Everything seems to be coming together quite nicely at the moment. My Beehaus is in transit and finally my nucleus of bees is almost ready. Fortunately there has been another little delay and so hopefully I will have the new hive set up in time for their arrival but based on what I saw in that inspection yesterday, I am also pretty confident that the jar of honey may be possible from that hive alone as they were going like the clappers.

I am really looking forward to seeing what all the fuss is about with regard to the Beehaus. It has received a lot of press and there have been lots of positive reviews and some not so positive. However, being a complete beginner I am going to be very interested to see which hive I get along with best.

I find it quite appealing that there are two hives lying side by side in the Beehaus; this seems to be one of the major positives about the design. I am slightly concerned that the old-fashioned method of collecting a swarm by giving your box to a swarm-catcher like Richard would be difficult with the Beehaus. It would be quite funny seeing them trying to catch the swarm in a brightly coloured freezer.

JUNE 17

I popped up to the hive this evening and, yet again, the bees have gone through another 3 litres of sugar syrup. I will pop up again tomorrow morning to feed them some more as it seems they are taking it in at a tremendous rate, which is a great sign. I was also pleased as, for the first time in several visits, I didn't get chased away from the hive; they generally don't seem to like me getting close, which is fun when quickly topping up the feed with a large silver saucepan. I can imagine it being very funny for my neighbours to see me scampering away from a beehive while swinging a shiny silver saucepan around my head.

Tomorrow the Beehaus arrives, which is exciting. Last time I was a little nervous about the hive arriving. I didn't have a clue about anything and certainly not the construction of a beehive. At least this time around I have a lot more of an idea and also I know that the beehive comes almost ready-made. Someone told me today that they see the Beehaus as an oversized and expensive yoghurt pot! I will be fascinated to see if I have the same opinion when I receive mine but, then again, this did come from a 'natural' beekeeper and so I suppose I can understand. A natural beekeeper tends to believe in minimal intervention, i.e. few inspections and no chemical treatment for diseases. They also use what are called 'top bar hives' which, from my understanding, are based on hives used in Africa that don't have any wax foundation, leaving the bees to get on with their own thing rather than being manipulated.

I have to say, given my time again I would have loved to know more about this form of beekeeping as I do like the ethos behind it. However, I have spent enough time and money this year, Jo has put up with enough and I am not sure Farmer Ray could tolerate another hive! I would like to try it next year though, it sounds interesting.

Anyway, the Beehaus arrives tomorrow and I've been told I can go and pick up my nucleus next week. All fitting together quite nicely finally, isn't it?

JUNE 18

It arrived today. Again, I felt a little bit like a child on Christmas morning though I was waiting for the deliveryman rather than Father Christmas. It arrived at about 11 a.m. in three huge boxes and immediately I went out to have a look.

Fortunately it was lovely outside and so I grabbed a knife from the kitchen and leaped out of the front door, much to the amusement of the Polish delivery driver as she was walking back through our gate. I was like an opening machine and I reckon health and safety officers would have had a field day with my knife-wielding technique. I started with the smallest box and there were bits of cardboard everywhere as I ripped through it. Gradually I organised it a little better and I had plastic bits in a heap on one side of me and cardboard on the other.

I am not sure what I really expected when I opened up the boxes for the first time but part of me expected a fully built and functional beehive. Instead I got a whole lot of plastic, which I have to put together at some point this weekend. It didn't look too complicated when I looked at the instructions but I think I expected it to be a little bit simpler than this. Whatever happens, my bees are arriving next week so I had better get a wriggle on.

I do admit my decision to opt for white rather than a bright colour has only accentuated the freezer type feel to it. Either way, I look forward to getting it built and in situ.

Tomorrow is my second inspection check and I have a few concerns, which is rare as generally I have very few concerns in life. It seems, however, that both beekeeping and gardening really occupy every thought. My first concern is that tomorrow I will be doing it well and truly on my own with no Richard this time to talk to and calm my nerves. I suppose I will take my video camera with me to talk to instead and hopefully that might help.

Secondly, I have concerns about the feeding of my bees. I have been feeding them with gallons of sugar syrup – they have practically eaten all of the sugar from my local Sainsbury's – but I am unsure when I need to stop. My mind was telling me that they should draw out the entire comb in the brood box first and then perhaps I should remove the feed. This has been confirmed by my great bee friend Adam, who mentioned that you should feed them until you need to put the super on top. I will have a look at the frames tomorrow and if they have drawn the entire comb, I will pop one on top. Exciting!

Lastly, I am worried about the closed floor that Adam put on. As I now know, when housing a swarm you should put a solid floor on the hive's open-mesh floor to give them a little more darkness while they are getting settled – which mine now appear to be.

What I am concerned about is the changing of the floors and how I do this on my own, especially as Adam screwed this floor to the hive. I assume I have to remove the screws first from each side and that is where I get a slight pang of fear. I can't quite imagine picking up the hive filled with bees, placing it on the ground away from the stand, taking off the closed floor, replacing it with the open floor and then lifting the hive back on top. I just find the thought of this quite scary. Surely this is going to make them incredibly angry and I will find myself running towards the pond quicker than I can say beehive. How on earth do I swap the floors with just me for company and a herd of slightly prickly bees? More planning is needed before I do this, methinks!

JUNE 19

Crikey, the day before Father's Day, which has taken on a whole new meaning since I became one. Not only is it a special day but I also think fathers need it simply to rest after the day before Father's Day. I had so many jobs to do today; put up shelves, put up pictures, fix cupboard doors, mow the lawn, plant out various plants and dig over the allotment. OK the last few were self-induced but I'm not so sure about the first lot of chores. I am sure my workload multiplied today just so that I could take the day off tomorrow. However, I did accomplish another hive inspection unaided. I have to say it is quite a different animal doing it on your own and you feel a lot more exposed without a shoulder to lean on. It all started with a very humorous lighting of the smoker and the afternoon took quite an unexpected turn, involving a tin of paint and a bee...

I was only thinking as I went to light the smoker that I hadn't had any trouble lighting the thing and that I didn't know what all the fuss was about. As a result of then going through about nine matches with no success and finding that I was down to the last one in the box, I started to reassess my initial cockiness. Allowing the paper and cardboard to catch light to such an extent that you can throw a handful of the most delicious smelling cocoa shells on top to produce the smoke is not as easy as it had been before. This was fate – but fortunately the last match caught and I was away, confident that my smoke looked good and I would have no more troubles... little did I know.

I was slightly apprehensive and interested to see what the bees were like again. Would they be as feisty? At least this time I knew my way around – it sounds funny but the first time you deal with a hive 'in anger' is a strange one. Small things like knowing how heavy the frames would be and how much pressure I would need to apply with the hive tool to separate the propolis and, ultimately, knowing the temperament of the bees were factors that I now knew a little more about.

I tried a few different techniques of smoking the hive this time as recommended by a good bee friend, Bernie (@thechoirboy on Twitter). He stated I should smoke the entrance as well before going in. Having left them for a couple of minutes I lifted the roof and as Bernie had predicted there were none on the crown board, not as many 'hangers on' and the kamikaze lot didn't start up immediately either. Thanks, Bernie.

I felt pretty confident going into this hive inspection as my smoke looked fantastic (if you are a beekeeper you will know what I mean by this). The smoke gently escapes the smoker in lovely wisps and folds like a cloud. It hangs around like smoke from a cigar; thick, white and just hanging there in the air. It really is a lovely sight.

It was evident a few frames hadn't been drawn out therefore I decided to keep feeding them a little while longer. This means that I won't be adding a super this week, which is a shame as I would love to see if I could get this one jar of honey. Apparently a beekeeper will add a super when the brood box is filled with bees and all frames have been drawn out. This is the trigger for adding the super and it means that the bees are ready to start drawing out the super to deposit the honey inside. At the end of the season the beekeeper will then remove the super, which is where the excess honey is stored. I have this nagging feeling that I should be replacing the floor now that the colony is established but I still have my reservations on how to do it.

As I worked my way through the hive I started to feel a little bit more in control than I did last week, though I was still not 100 per cent comfortable. I am conscious that I need to monitor what I see and record how many frames have bees on, how many have brood, how many have stores, etc. etc. However, my mind just tends to become a blur. I am concentrating so much on (a) finding the queen and (b) being careful not to hurt or maim any bees that I find while counting the frames and retaining the vital information. It's surprisingly difficult. I think this is perhaps a skill that you learn with experience or perhaps one that comes with swigging a little whiskey beforehand to calm the nerves. Though maybe that would aggravate them more, as I hear they don't like the smell of alcohol.

It was all looking good; I continued smoking, as you are encouraged to do (though not always smoking the bees). I then got side-tracked as my attention was drawn to simply getting through the inspection. My smoker started to suffer and I realised that nothing was coming out. I struggle to describe what happened next as it was genuinely quite unique. It was as if the bees had sensed that my defences were down! Let me first say that I was told that bees do not, under any circumstances, like you even blowing on them (I had this reinforced earlier today when I let out a sigh of nervousness while looking closely at a frame and promptly saw about two hundred bees desert the frame they were on and head straight for my veil!). Here I was essentially blowing a mass of hot air on them with little or no smoke. If you can imagine a scene from *The Lord of the Rings* with the orcs coming out of the woods; a few at first, and then, sensing the path is clear, they literally stream out of all possible gaps. This description is not far from what I witnessed today. There were a few on the tops of the frames and then suddenly as they realised (a) I had run out of smoke and (b) I had just annoyed a few of them with a hurricane-force wind directed straight at them, they all started filing out of the frames and on to the top. The buzzing sounds started to increase, which happened to be in direct correlation with my heart rate. Keeping calm, I just had to finish off the final few frames.

I realised that I had again not seen the queen. Laura, Suzy's daughter, had mentioned to me during the week that I should be able to see Cleopatra quite easily, as the blue spot on her thorax is very noticeable. Therefore I can only assume that I have Nefertiti (incidentally they have now named their new queen Hatshepsut) but I just can't spot her. I am sure this is a combination of my eyesight being akin to that of a bat and also my lack of experience.

Anyway, I rather anxiously put the hive back together again and tried to estimate what I had seen. I am pretty sure that I have bees covering at least seven frames and have eggs and larvae in at least six of those. I remember seeing sealed brood on about three frames which means I should have an interesting inspection next week. My colony will have multiplied considerably, I would have thought. I am looking forward to that. I learned one major lesson today, though: keep your smoker smoking!

Many people describe beekeeping as relaxing and totally absorbing. The latter I can agree with, but the former is yet to be accomplished. I am sure it will be in time but I am still finding it pretty scary each time I enter the hive. Even so, I felt pretty pleased with myself as I left the hive, followed by about twenty bees wanting to make sure I was going.

In the afternoon, among many of my pre-Father's Day jobs, I sneaked in one of my own, to paint a couple of doors. My main target was the Man Shed door. It had obviously been painted before but it looked more than a little distressed – as if a bear had been using it as a scratching post for the last year. Therefore I went to paint it; but it was about 10 feet from the hive I had inspected only an hour earlier.

I thought an hour would be plenty of time for the bees to have calmed down but as I started painting, I got attacked; they must have remembered me. The thing about bee attacks (that does sound a little extreme, but at the time that is what it feels like) is they are rather sudden, you never know it is about to happen. Therefore, in this instant, as I had a paintbrush in one hand and a tin of paint in the other, a bee made a beeline for my ear and kept on flying at me.

A saucepan in your hand is one thing but a full tin of paint and a paintbrush is quite another. Off I went (I have a planned escape route now) but this time to dire consequences. I was desperately trying to keep the tin of paint still while waving foolishly with my other hand complete with paint-covered paintbrush. Stupidly

I should have remembered that bees are attracted by movement but this all goes out of the window when you have no veil for protection! It was also pretty stupid to be waving a paintbrush around. I was covered. Paint in my hair, paint on my clothing, paint on the mangetout and pumpkins. Paint everywhere. At least my escape route worked and I was unscathed. Suffice to say, that 'man's job' can wait a while longer!

I therefore got on with building my fridge-freezer of a Beehaus. Despite there being more pieces than I initially thought, I have to say it was a joy to put together given the glorious weather. It was like constructing a Lego kit and it all slotted in place very easily.

Jo, Sebastian and I were just lazing around in the garden as I took my time over building it. A couple of cold beers were shared and the garden enjoyed. There is something about lying on the lawn on a summer's day, shoes and socks off with your toes enjoying their new-found freedom amongst the newly mown grass, and sipping an ice-cold beer. It really was bliss, though it was just slightly strange to have this sparkly Beehaus next to us. Still, once it was built, I thought it looked fantastic and couldn't wait to have it next to my traditional National hive. I will move the Beehaus into position tomorrow I think, and enjoy today for what it is. Loads of bees are flying around, incidentally. I wonder if any of them are mine.

JUNE 23

So... you can get stung in the eye.

I had always thought this was a fallacy as part of my initiation but seemingly not, as I have found out at my association training tonight.

It was the most glorious evening to be beekeeping and was my first evening at a new apiary for the Reigate Beekeepers. They have been in the process of moving to a new apiary for some time now and it was worth all the effort they have put into it. It was a stunningly beautiful sight. There are about ten hives now in total, all set out in a large circle measuring about thirty metres wide, which makes for quite a spectacle and a great training ground as there is ample space to move around. It is in the most glorious grounds surrounded by trees and meadows. There really couldn't be a better setting.

I haven't been for a couple of weeks and so it was really nice to meet up and get down to business with our regular hives. It was refreshing to deal with some calm bees. They seemed so relaxed compared to mine and it was a joy to handle them; I even considered picking one of them up but that thought quickly passed.

What was interesting this evening was the role we play as beekeepers and I hadn't thought it would be so philosophical. To cut a long story short we had to decide whether or not to kill off a queen as the colony had already been artificially swarmed (a long and detailed manoeuvre which I won't even attempt to explain as I am not sure I fully understand it yet) but her remaining daughters had raised a new queen cell so they obviously weren't happy with her.

It felt like we were playing God with this hive, which I wasn't entirely comfortable with. Her survival rested on a vote which, with my vote, spared her and we were of the mentality to let the bees sort it out; far better in my opinion.

Casually, towards the end of the session, Richard decided to slip into conversation that last week he was stung on the eye. I couldn't believe it – this is the man who helped me film my first inspection visit and is always such a reassuring presence! Yet there it was on his eyelid, a couple of millimetres from his eye, a small scab.

Apparently, last week at this apiary session, he had been clearing away having just completed a hive inspection, a good fifty metres from the hives, and a bee took umbrage against him and another beekeeper. It flew straight at him, apparently, and then the next thing he knew he was grabbing his eye in serious pain. The other beekeeper standing next to him was in fact Tom, who first told me that you can get stung in the eye. With this being the second time it has happened near Tom, I wonder how many people will stand near him again! However, knowing him a little better now, I bet he was secretly pleased as it meant he could recount this story once more with some naïve newbie beekeepers.

In the end after some immediate medical attention at the apiary, Richard didn't go to hospital but he did take an antihistamine tablet. Despite that, his eye apparently swelled up considerably and was practically closed. It was painful for a day but then it was OK. Needless to say, a lucky escape – if he'd been stung a couple of millimetres higher, then it would have been a whole different story. This event combined with my recent close shave up at the hive suggests the need for far more care around the bees than I was giving them at the moment. Lesson learned as far as I was concerned.

The evening finished with a pint at a new watering hole close to the new apiary location and we saw the sunset overlooking a beautiful lake. Fabulous – I could get used to this.

JUNE 25

'The call' came yesterday. I had already left home for my mother-in-law's when my phone rang. 'Do you want to pick up your bees tomorrow afternoon?' asked a very well-spoken gentleman. 'Yes, yes, yes!' went my heart, 'No, no, no!' went my head. The Beehaus had arrived but it was lying on our lawn and I had to get this white, freezer-type contraption up and in position at the allotment. Not an easy feat given its design. Not to mention I had yet to build any of the frames needed to put in it. How on earth was I going to do this from my mother-in-law's house?

'No, no, no!' was shouting louder in my head and yet I heard my mouth open and utter the immortal words, 'Yes, I would love to pop by tomorrow to pick up the bees.' No, no, no, no, no! What on earth had I agreed to? However, I had messed this guy around enough and therefore set about resolving the plan in my head last night. I was struggling to think how exactly I was going to work it all out. I came to the conclusion finally that there was no way I could do all of it and I would just have to give it my best shot. Not a nice thought to go to bed on. I had again not planned it all very well.

When I awoke it became clear immediately that it was going to be another very, very hot day. Hose-pipe bans are being discussed which is quite incredible given the wet winter. My greatest concern, however, was my nice black car which wouldn't make the journey with bees any easier. As I wasn't transporting a complete hive this time I decided on our saloon car and I was hoping that I could pop the nucleus in the boot, far away from where I was sitting. There is also no way that bees could escape the boot either which was a bonus while driving. Anyway, I set off and arrived in good time at the most glorious of homes, one I only ever expected to see in those expensive property magazines. As I drove up the pebbled driveway, this beautiful, old, wisteria-laden house emerged from behind the trees. It looked ever so peaceful and as I slowed almost to a stop, the small wooden door opened. An elderly gentleman, who I took to be Alastair, emerged. He was slightly hunched over to get his tall frame through an impossibly small door. As I left the car we exchanged pleasantries and shook hands. I didn't dare say to him that I hadn't really got anything prepared back at home.

Money exchanged hands and the correct position in the car was decided. When taking the swarm the boot was really the only place we could put it in, but this time Alastair suggested the back seat complete with seatbelt to keep it secure; sadly ruining my plans of keeping them as far away from me as possible. Generally, when you are buying a nucleus, you will be given it in what is termed a nucleus box which is still made of wood but is about half the size of a hive, and therefore far lighter. I have seen some online that look like very high quality wooden boxes but what was in front of me seemed more the norm. It was a plywood box with a very thin wooden lid, complete with a hole fitted with gauze for ventilation. It was all held together with masking tape which I thought a little strange. This also meant it was a little bit more unstable and hence the need for a seatbelt. The chosen position made it a little bit more daunting as they were right behind me and I could really hear the buzzing but I set off anyhow. I said my goodbyes with one last statement from Alastair: 'Better get them out as soon as possible because of the heat, you wouldn't want them to go into meltdown – heaven knows what they will do.' Gulp.

I was making quite good progress around the winding roads of Surrey until I heard a sudden pop and a hiss. I simply couldn't believe that in the midst of transporting several thousand bees, I was having my first ever tyre blowout. To make matters worse it was really heating up outside. Me and my several thousand passengers slowly rolled along, desperate to find somewhere to pull over.

I managed to pull in to a layby almost immediately in front of a couple of gentlemen in a lorry. I got out as these two giants of men clambered down from the cab to say 'O'wight mate, don't look too good does it?' Having agreed with them and said a quick hello I got out the spare replacement wheel, which is a sorry-looking thing half the size of a normal wheel and restricts you to travelling at the speed of a snail, while they inspected the problem. Seriously, these guys were huge, wearing wife-beater vests with muscles bulging and complete with tattoos on every inch of their body. They were very helpful, however, until I mentioned that I had to get the bees out, which is when their tone changed to one of concern. I got the bees out of the car as the two guys backed off a little ('I'm allergic,' I heard one of them explain to the other) and put them in the shade to cool down.

Alastair's closing comments were ringing in my head as I got on with the task of changing the tyre. The guys soon left as they realised that I would be OK. That feeling of satisfaction when I realised I knew what I was doing with the jack, being able to release the wheel nuts with the 'locking wheel nut' and getting the wheel off had a strange effect on me. With the precious cargo to my right buzzing away, it was a very masculine moment, one even greater than chopping wood or lighting a bonfire! Still, I cannot imagine what those guys will be saying down the pub this evening about helping some random bloke change a tyre complete with a hive of bees in the back seat.

Once the tyre was fixed, I got on my way and managed to make it home in one piece. Now came the first part of the plan I had concocted over the last twelve hours: the Beehaus move to the allotment. I removed the bees from the car again to give them some fresh air and placed them in the shade before looking at the Beehaus again. It looked heavy and awkward and sadly Jo had the larger estate car so I couldn't simply transport it up with that. However, I realised if I broke it down into two journeys it would be OK and so firstly I took the roof, balancing the supers on top. I went from what I call the African style, i.e. balancing it on my head, to bear-hugging it looking through the open-mesh floor on the bottom to aid direction. Either way, it was not easy and a black mark against the Beehaus but I'm sure not all moves are done at this level of haste.

I got it all set up and then did exactly as I was told. Thankfully, I didn't have to build all the frames immediately as I was told to put the nucleus on top of the newly set up Beehaus for a couple of days and let them out before moving them into the hive. The bee suit was on and I tentatively pulled out the bung to let the bees free. And nothing happened.

It was as if they were unsure, but about ten seconds later a little head popped out, quickly followed by another and then they streamed out into the big wide world. Lovely.

I shall pop up tomorrow to see them into the Beehaus proper, complete with the frames that I still have to build early tomorrow morning. What a lovely experience though, and despite being more expensive than a swarm, it's far more civilised in the way that you retrieve them, and it is also good to know your bees have come from a good home. Ultimately I suppose this is the most important aspect of buying a nucleus because you are buying them from a known source rather than just picking up someone's swarm from somewhere.

All in all I am pleased that I have now experienced both methods of obtaining bees; I only wish they had both been a little earlier so that I could really take advantage of a whole summer for them to make me some honey. I hope I am not too late.

JUNE 26

Good God, I am not sure I can recommend making frames at 6 a.m. in the morning. Usually I just about get away with digging holes or watering at this time of the morning as little can really go wrong. Using a small hammer is certainly not the best thing to be doing before the early morning coffee and my thumbs certainly know about it.

This weekend feels like quite an important one for my beekeeping career. Today will be the day that I introduce my girls to their new home and tomorrow will be the day that I introduce my traditional hive to the open-mesh floor that I have been stalling on transitioning to. And I may have to add a super on top as well, which is exciting.

Once the frame-building was out of the way to the delight of my thumbs, I got on with the rest of the day. I heeded Alastair's fine advice of leaving the bees till evening to deal with their big move. Therefore Sebastian and I went to play football together and we all then spent a very pleasurable afternoon basking in the unbelievable temperatures while seeing my old university friends for a barbecue. What a nice way to spend the day, knowing that later I was due to dabble in the unknown...

No wonder it was nerve-wracking – I simply didn't know what I was about to do. I was used to using traditional hives, and the Beehaus was something completely different, requiring a different way of working. Most of this was due to its size but it also makes you work a certain way. Whereas with the National hive I can work the frames from behind the entrance, the Beehaus forces you to work from the side which I suspect will feel a little awkward. This along with the fact that I was about to perform a manoeuvre I had only seen on YouTube didn't help matters – I would be flying by the seat of my pants! Not a good thing.

All day, in the quiet moments, I was working out exactly what I had to do and was trying to formulate a plan. I hadn't really had time to play around with the Beehaus since I built it. I still hadn't put any frames in or any cover boards on. I obviously haven't learned from my previous planning issues, or maybe it is just that beekeeping is not quite suited to those who have a child, a pregnant wife and a full-time job. I would like to think it is the former, personally.

Evening arrived and Sebastian was tucked up nicely in bed. Tentatively I made my way up to the allotment, reworking the plan in my head. I had made the feed in advance to add on at the end so I knew this was all OK but I was still worried about everything else.

I got to the hive and sure enough there was no activity outside so most must have been inside, which was a good sign. I laid out all the equipment I needed on the ground within easy reach and went to light the smoker. It lit first time. I smoked the entrance and the three or four bees poking their heads out dashed inside; I quickly popped in the bung and felt quite satisfied that part one of the plan had come off. I could now move the nucleus and prepare the hive for them without bees flying all over the place and wondering what was happening with their home; and more importantly, I didn't have to rush.

I got all the frames in place (the nucleus only comes with five frames and so I had to make sure I had more in the hive to fill it up and give them room to expand) and then started on the bees. It is funny; this was only the third time I was dealing with bees on my own and I felt so much more relaxed about it now. In fact that calmness that beekeepers talk about was starting to take effect. However, I had to be on top of my game here and so I continued.

After giving the bees a puff through the gauze hole in the lid and then struggling for a bit with the sticky tape holding it all together, the lid came off. They were so calm compared to what I was used to with the other hive. It was lovely, and so I started to slowly remove the first frame. It was only at that point that I realised I had done something incredibly stupid – I had left the hive tool in the shed! I am not sure how as I felt so organised, everything was laid out similar to the way that you lay your clothes out on the bed before you go on holiday. This was comparable to forgetting your pants. I had left myself open to the elements but I just had to get on with it.

Fortunately, the frames came out relatively easily. Slowly I inspected each frame and placed them one by one into the hive. It was all rather nice but I was desperate to find the queen. Nothing on the first four frames and yet I knew this queen was marked with a small dab of paint on her thorax to make her easier to spot;

a common thing to do amongst beekeepers. I was beginning to think that it was my eyesight that was the problem.

However, on the final frame there she was. It was like a eureka moment; I felt so chuffed as this was the first time I had ever found my own queen and quite unexpectedly I shouted out 'Hey there, Queenie!' I was so pleased and yet also relieved as I knew that I did have the ability to spot the queen. My God, I thought, she is massive.

Suffice to say, I got them all into the hive safely and put all the bits back in the right places. My plan came together and I felt rather pleased with myself. I had worked out how to deal with a hive that no one locally has used and secondly I had found the queen.

On this lovely Saturday evening, therefore, I am feeling on top of the world – till tomorrow, and my attempt to change the floors on the other hive; that will take planning to a whole new level.

JUNE 27

There is something quite satisfying about going up to a hive after putting a new set of bees inside and seeing them flying around quite happily. It is nice knowing that I must have done something right for them as they seem to be carrying out their orientation flights to familiarise themselves with local landmarks, and I have seen a couple of them flying into the hive with pollen attached to their legs.

There is one slight concern, however; I am not sure if the Beehaus is 'bee-tight'. They seem to be sneaking in somewhere other than the entrance as there is a lot of activity at one of the gaps between the supers. I have closed up the gap to the best of my abilities and will see if it makes a difference. I must check this later.

The main job today was to finally change the floors of the traditional hive, having put it off for weeks. I had concocted a plan over a rather strong coffee this morning and had now got everything ready. Equipment was lined up beside the hive like I was preparing for major surgery. Smoker lit, hive tool ready and screwdriver in hand ready to unscrew and detach the old floor.

I unscrewed the metal joints holding the floor and brood box together and then smoked the entrance. Slowly I lifted the complete hive off the stand and placed it on the floor beside me. A few of the bees were flying around but it was OK so far and they didn't seem too rattled.

I placed the open-mesh floor on the hive stand and started to tease apart the hive and closed floor beside me. This was all in readiness to lift the hive back onto the stand. This was the bit I was afraid of because essentially I was taking the floor away from their home. Imagine the human equivalent of this and seeing the floor of your house disappear while some grubby gloved fingers gripped the walls to lift the whole thing up in the air. It would be a little unnerving but here I was doing it to probably 20,000 bees.

I had only lifted the hive a couple of inches off the old floor when it happened. My first sting followed quickly by my second, third and fourth.

Initially I thought I had been stung by the nettles around the hive. There were short stabs of pain around my ankle. As I continued to manoeuvre the hive it dawned on me what was happening. I put the hive down on the floor a lot more carefully than I thought I would have done considering this new development. However, the thought of dropping a complete hive and annoying the whole colony was simply not an option.

I bent down and tried to flick the bees off from around my sock area but realised very quickly that using a hive tool in a swinging action close to the major arteries of my foot wasn't the best plan. Therefore I resorted to the fingers and quickly dispersed the stinging insects. Reviewing the situation now I had been pretty stupid. Not only had I put the entrance of the hive right by my ankles but in order to prevent bees from crawling up my trouser leg I had done the other fashionable thing of tucking my trousers into my socks. Not only did I look really cool but I left an open invitation for them to attack my ankles. Silly, isn't it – by trying to stop bees crawling up my leg I had left my ankle exposed. Maybe next year I need to either buy a full suit or purchase some welly boots.

All in all it was over very quickly and I have to say I was quite glad that I have now been stung, especially as I wasn't expecting it. I figured my hand or fingers would get stung first. At least I know what the feeling is like and I can be doubly pleased that I didn't get stung in the eye. The stings hurt far less than I had expected and there doesn't seem to be any swelling so that must be a good thing.

Anyway, all done and the hive was now on its correct floor but despite being stung I now needed to carry out the inspection on the Beehaus.

Fresh from the confidence I had gained in spotting the marked queen yesterday in the nucleus, I found the unmarked queen today! Even though I had seen them in the practice sessions I am still amazed at how big they are.

Having done a few inspections now, I have to say with every opening of the hive my nervousness abates ever so slightly and I am feeling a little bit more comfortable around the bees. Being stung earlier has helped as well; as arguably nothing can be much worse than being stung four times, aside from being stung five times I suppose.

As I was working through the hive I started to realise that I was viewing each frame differently from past inspections. It was almost as if I had a checklist going on in my head. On each frame I started by checking if there were any larvae, which gave me the confidence that I had a queen. I then moved on to whether I could see stores and any signs of disease. Finally I would just check the bees to see if I could see the queen as I scanned the frame left to right and top to bottom. After the third frame I realised I was doing this instinctively and again it made me think how enjoyable the experience actually is. It was, however, the time that I also realised how hot it was inside the bee suit. I may as well have simply wrapped myself up in a dustbin bag and stood in the sun. I was baking and I could feel sweat droplets dripping onto my nose.

Anyway, the queen was found, the hive floors changed and I left a very happy novice beekeeper. This was especially the case as I put on the honey super as well. I felt that the brood box was sufficiently full to chance my luck. Hopefully this will incentivise them to move upwards, build some frames and pop in some lovely golden honey. Perhaps this will go some way to obtaining the jar of honey I so crave this year.

JUNE 28

I am feeling quite old today. Imagine the scene. I had to get the bus into work as my car was getting its tyres changed (after my blowout incident). Once the bus got going I realised that I was on the side that the sun was streaming through. I moved sides and it was at that point I felt old. I actually moved in the interests of a comfortable journey.

Matters got worse. As the journey continued, the schoolchildren joined me on their way to school – with spots coming out of every inch of skin, hair down to their ankles and ties with the loosest definition of a Windsor or half nelson knot I had ever seen. There I was minding my own business, dressed for work in a smart shirt and well-tied knot and felt I should catch up on some reading. Out popped a copy of *Beecraft* – Britain's best-selling bee magazine. I started reading it and realised that some of the kids had brushed aside their ankle length hair just a little bit so that one beady eye was looking at my reading material. These brief looks turned into stares which were followed by elbowing and a bit of pointing. The receiver of the elbow would then turn, wipe their fringe in a certain direction and then also stare in a way only teenagers can.

I was innocently reading a magazine with a nice photo of a bumblebee on the front and being stared at by a bunch of teenagers. OK, so it wasn't the latest edition of *Heat* or *GQ* but I thought I was pretty cool. Then it dawned on me. Me and my mates used to be those spotty, long-haired teens but the recipients of our stares were people reading trainspotting magazines. I couldn't help but laugh back then at these anorak-wearing, bespectacled human beings and here I was, the modern-day equivalent. I felt mortified but, as I am sure the trainspotters did, I buried my head in my magazine.

Once I had returned from work, I went up to the Beehaus to check on the hive to see if it was bee-tight and thankfully it all seemed in good shape. I also went up to give them some more feed but upon opening the hive I discovered they had hardly even taken any – in fact I would go so far as to say they hadn't yet found it despite my efforts of introducing it to them. As a result they weren't drawing out the frames particularly quickly so I dripped a little bit more down the tube to incentivise them. Hopefully they will get started pretty soon.

It has been 29 degrees Celsius today again, which hasn't made it a particularly comfortable time, especially with my stings having been covered with sweaty socks all day. My ankle became very itchy today for the first time and was a little bit inflamed and swollen.

On a separate note, I had my first spuds this weekend. I dug up some of the swift earlies which I had planted several months ago. A little bit of mint while cooking followed by 'accidentally' too much butter, salt and pepper and all was delicious. I will also say that, despite the odd few, today was the first harvest of the mangetout and broad beans. I made a lovely salad complete with pepper, potato, rice and coriander, and all was lovely. So satisfying after all the hard work I put in throughout spring. Wonderful.

JULY 2

I haven't written much this week because, quite simply, my hands have been too busy scratching away at my left ankle. The stings have developed through a variety of stages throughout the week which has been most interesting. On Sunday, nothing really happened, which was disappointing. I was expecting or rather secretly hoping to see some swelling immediately after the stings and maybe something like the elephant man's leg but I got nothing – no evidence of the stings, not even a red mark. I had walked back to Jo hoping for sympathy and had nothing to show for it.

Monday arrived and I remember jumping out of bed and looking down at the ankle – still no elephant man-type symptoms or even any redness. However, as the day wore on, I found myself scratching the area of the stings and it just got progressively worse.

I woke up on the Tuesday after a very unsettled night. I think that I must have been trying to scratch the inflamed ankle with my other foot as I do remember kicking Jo several times during the night. I was in serious trouble with her in the morning for precisely that reason. I thought I had a decent excuse though, especially as my ankle was now nicely red and there was finally evidence of the stings. Sadly my defence was pretty much dismissed. Kicking your wife while asleep because of a few bee stings is obviously not recommended.

Ever since the pleasure of actually proving to people on Tuesday morning that I'd been stung I have been feeling pretty fed up with the matter to be honest. They have been incredibly itchy and though the swelling went down yesterday, Thursday, they are still quite uncomfortable. I have a feeling that by Sunday they will have calmed down completely just in time for another inspection. I wonder if it is all contrived by the bees. They deliberately pump in enough venom to keep the sting uncomfortable for a week, just so that others will sting you on the next inspection, keeping the level of comfort at such a point that you eventually decide bothering them is a bad idea!

JULY 3

Today was a strange day. I woke up feeling quite excited following on from last week's successful inspections. Spotting the two queens gave me a real confidence boost and then there was my growing feeling of calm as I inspected the hives, something I hadn't felt before. Today, however, was slightly different. I got just over halfway through the inspection on the traditional hive and then felt that I couldn't complete it.

It was funny as I had the game plan mapped out in my head today. I wasn't going to open the Beehaus, having decided to leave it another week to allow the queen to get herself up and running. I think this is normally accepted protocol when you have just introduced a nucleus to a new hive. Therefore I had only the one hive to inspect, the traditional hive which held the ever-growing if slightly feisty colony. It should have been quite straightforward.

I think the trouble started about an hour earlier. I was doing some work up at the allotment near the hive and had spent about five minutes observing the hive from a distance. It was incredible and the only way I can describe it is like a really busy road at rush hour – a constant stream of traffic going both ways, invariably faster than the speed limit but, for the most part, staying in the right lane and all working in harmony. It was incredible; I was mesmerised. I could clearly see the constant stream of bees going in and out. It was slightly silhouetted with the Gatwick Airport planes coming in to land in the distance and yet the two flight paths couldn't be more different. My thoughts went to the impending inspection and I was suddenly a little bit psyched out. My God they look busy, there must be a lot going on in there.

Come the hive inspection, it started well; I got my smoker lit first time, bee suit on, gloves on, head engaged and hive tool ready. It was the first inspection I was doing with a super on top of the hive; something I hadn't encountered before. Anyway, I got on with it and it seemed to work well as the smoke at the hive entrance did its job.

I lifted off the super and was immediately aware of the increase in bees that I saw under the queen excluder (this separates the brood box from the super and keeps the queen in the brood box so that she doesn't lay eggs where honey is stored) and already in the super going about their business. They were busy drawing out frames in the super which was a good sign. It was evident that there were a good load of bees all over the frames in the brood box. This immediately pleased me as I knew it meant a lot of the brood must have hatched last week. Off came the queen excluder and I got stuck in. My mind was racing with excitement and any previous concerns went out of the window.

After the second frame I started to see freshly laid eggs, which I was pleased about, but then my right glove got stuck on the propolis and the thumb of my glove ripped clean off, which tells you just how sticky this stuff is! Immediately I got a little bit concerned. It is amazing what a tenth of a millimetre of latex glove does for your confidence. I was then immediately aware of just how hot it was outside – the outside temperature must have been at least 28 degrees Celsius – and that again I had sweat dripping down over my nose, not a pleasant feeling knowing that I couldn't really wipe it away.

I carried on regardless but was acutely aware of the bees being a little feisty. These girls have always been a little bit feisty but this time there were more of them! They were dive-bombing my veil like kamikaze pilots and I could deal with the ones at the front but when they were going for the back of the head that was a different story.

I got to the seventh frame but was becoming increasingly uncomfortable. I was hot and bothered and the bees were all over me, especially on my gloves. Every time I removed a new frame more bees piled out and it was as if they were making a decision every time: do we stay here or would it be more fun to join the kamikazes or go for that exposed finger? As a result of my increasing nervousness and generally not feeling happy about the whole situation, I took the decision to finish the inspection early. I had seen the eggs and so I wasn't too concerned but I just wasn't happy with the situation and there seemed so many more bees than normal.

As I closed up, I quickly took a look at the super and made sure the individual frames were doing all right. I had split my super into half 'cut comb' (so I could make cut comb honey) and half normal frames. The cut comb frames basically have no wire going through the wax and the result of this and my bad workmanship of fitting these particular strips was that the wax had worked its way loose and was hanging loose in the frame. They would be useless and would have to be replaced at some point. Damn; a bad ending to a bad inspection.

Sitting down now, glass of white wine in hand, reflecting, I am both pleased that I took the decision to stop the inspection but also slightly disappointed that I couldn't continue.

JULY 5

I am sitting outside having just tied in some tomato plants and am having a glass of wine: bliss. It has given me some time to reminisce about Saturday. Despite feeling a little bit of a wimp, upon reflection I think I did the right thing. If you don't feel right you should pull away from the inspection.

This morning I went up to the allotment on a daring mission to cut the grass around the hives. It might not sound much but imagine as a child approaching the house of the neighbourhood weirdo who you had been told ate children. You had been dared to knock on his door and then run away at the speed of sound. You probably knew deep down that he didn't eat children and so, with your heart pumping you would have done it but would have felt so unbelievably nervous all the same. That was the feeling I had this morning.

So I got up nice and early, hoping to miss the rush hour around the hive. The M25-type traffic I witnessed the other day would hopefully be several hours off yet and so I might be able to sneak in and get the job done. It was absolutely beautiful this morning and the field in front of the allotment looked majestic as it often does in the early mornings.

I went to the Man Shed, pulled out the hand pruners (not really sure what this contraption should be called but basically shears on long handles) and crept up to the hive. I was going rather quietly, careful not to draw attention to myself, and I have to say it did raise a question in my head. Can bees hear? If not then I was wasting my time being silent. I have a feeling that it is only really movement and vibration that they are aware of but I could not be too sure this morning.

I got started and to be honest it was a rather quick job and I was in and out of the area within five minutes. I had given the grass around the hive a good haircut and all was looking rather smart. I was quite pleased with myself and then decided that I would start on the loose branches around the hive as well – shows what a little bit of confidence does for you doesn't it? Anyway, this job was also done pretty quickly and all was good.

Afterwards I sidestepped over to the Beehaus to have a look at what the bees were up to. I lifted off the lid really delicately – I have to say the more times I use it the easier it becomes, quite practical really – and looked in. I was really pleased to see about fifteen bees in the feeder which was fifteen times as many bees as I had seen previously. It was evident that the feed had also been devoured so my intervention last time had obviously worked. With the crown board on it meant that I couldn't see into the actual hive but looking into the feeder gave me a good indication that things were going well.

Having left the Beehaus bees alone for a couple of weeks I am looking forward to doing a full inspection on Saturday rather than just looking longingly at the entrance, to see how they are getting on in this most controversial hive. I am quite enjoying looking at the two of them side by side. Even Farmer Ray pointed out that I had 'one of them modern hives' the other day – even the scary farmer is interested!

I put a couple of litres of feed in the feeder and left them to it. I was a much happier beekeeper on this very beautiful morning.

JULY 7

A little bit of confidence was restored today as I went to my evening with the Reigate Beekeepers and got my hands dirty once more. It is really uplifting to be around other, more experienced beekeepers who all state it is quite common to abandon a hive visit. Some, by the sound of it, have had it a lot worse than me!

The session started ominously with my gloves splitting once again but I just got on with it this time. The weather wasn't as hot as Sunday but the hives there seem a lot calmer than my own. It was a nice and gentle introduction back into the swing of things for me and a much-needed morale boost.

I am slightly concerned about what I might find when I carry out the inspection on the Beehaus as (a) they haven't taken too much of the sugar syrup, perhaps indicating they haven't been building much onto the wax foundation and (b) I know the hive is on a little bit of a wonk and it needs to be straightened up. The latter should be straightforward and hopefully the former should be OK too, as there is a very strong nectar flow at the moment, which will encourage them to build out the combs naturally.

Talking of nectar flow, when we finished the session we accidentally popped into the local pub for a drink and sat around discussing our bees. There was more talk about this strong nectar flow or honey flow – the terms seem to be interchangeable – but something else sparked my interest tonight and that was the discussion about extracting the honey. Up until now I have only dreamed about extracting a jar of honey but time is nearly upon us when this may become a reality; I should get a far better idea this weekend.

It became clear that extracting was not an easy process, was quite labour intensive and required some pretty technical equipment – despite some of the conversation being centred around DIY equipment, much to the hilarity of others with stories about mistakes made using this sort of equipment. I suddenly realised that within eight weeks I could be spinning some frames around in an extractor and looking at my own honey. Andrew, the person responsible for me joining Reigate Beekeepers to begin with, is also the 'equipment guy' and stated the association's extractors had already been reserved for a lot of July and August.

Once I have had a look over the hives this weekend to see if there has been any honey deposited, I must get a reservation in. I hope I am lucky enough to be in a position to be able to do this, as how else could I extract the stuff?

Incidentally, I also found out about ankle biters today. There seem to be some bees that go directly for the ankle and apparently stings there are particularly itchy. Don't I know it! Not sure Jo has forgiven me yet.

JULY 8

After last night's pub discussions, the enormity of the extraction process was preying on my mind. All the equipment needed and the fantastic game of dare you play with the bees when you take their honey super, not to mention the mess I will undoubtedly leave my kitchen. However, a chance meeting with a friend of mine in the coffee shop I go to every lunchtime and the suggestion I should watch how the Nepalese extract their honey has calmed me down considerably. I walked into my beloved coffee shop and there standing by the till was Steve, an old school friend of mine. Despite the fact he had far less hair on his head, which seemed to have fallen southwards to make the most glorious of beards, this couldn't disguise his cheeky grin which was instantly recognisable. We sat down and got talking about the past before catching up on what we are currently doing which inevitably led on to my beekeeping exploits. Having described what I was up to and the hopeful extraction of a jar of honey in the near future he couldn't wait to tell me about a recent TV programme he had watched.

The Nepalese honey hunters, as they are called, scale vertical 600-metre-high cliffs with vine ladders with very little protection. They don't get too close to the huge combs of bees and honey, made by the largest honeybee in existence, which is understandable now having seen them. They use a combination of baskets controlled by others 40-foot higher and a 20-foot-long pole with a knife embedded in the end. With amazing coordination, skill, dexterity, braveness and teamwork they cut away these honeycombs before lowering them into the ether below. This is where the rest of the team wait, whose original job it is to light the bonfires to smoke the bees.

It is truly awe-inspiring to watch (I was able to watch it online, fortunately) and puts my honey extraction of two hives, not even 600 feet from my house, into perspective.

However there is still the small matter of equipment. I think, on the basic research I have now done, that I need the following:

A honey extractor – though this will depend on how much honey I have to extract. If I don't get much there would be little point and I could probably manually extract it (note to self; must look at how you manually extract honey).

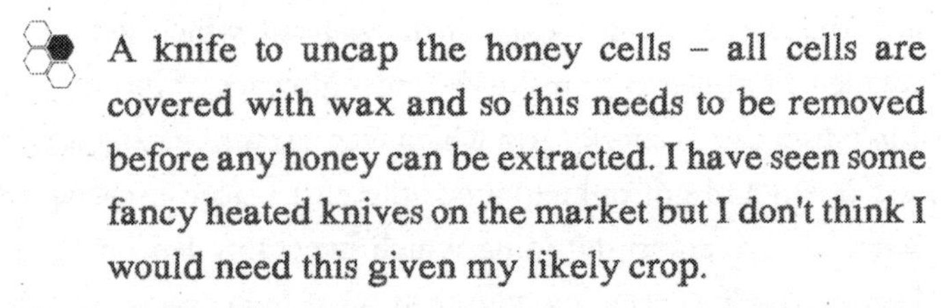

- A knife to uncap the honey cells – all cells are covered with wax and so this needs to be removed before any honey can be extracted. I have seen some fancy heated knives on the market but I don't think I would need this given my likely crop.
- A container to keep the uncapped cell cappings – it is apparently best to keep them separate as they would clog up the piece of equipment mentioned below.
- A filtering device – you can buy these great sieves to filter the honey but I am thinking muslin would do the trick because I am hardly likely to have too much honey. Something is needed though as apparently honey in the cells can contain remnants of bee wings and legs which I am not sure I want in my one jar of honey!
- A settling bottle/tank – apparently it is best to leave the honey for a while to allow air bubbles to rise to the surface. Otherwise your jar of honey will be full of bubbles. Again, I don't think I will be too fussy here but it might all be worth looking into.
- Lots of jars – if I am lucky.

So my preparation starts here. Let's hope the bees are able to fill a super for me to extract.

Something else I realised last night was that I had made another small mistake: though I had replaced my closed-mesh floor with an open-mesh floor, I hadn't actually opened it! What was I thinking? There is a small floor within the open-mesh floor design which is removed to allow air circulation or closed to catch debris – vital to assess any possible disease in your hive, especially varroa mites. For some reason I never removed it, which was a little stupid because in this unbelievably hot weather, it was essentially a closed hive. They must have been cooking!

Anyway, I quickly ran up to the hive in the twilight and got started. I then realised that I hadn't removed it since I painted it and so it was essentially glued shut. Out came the hive tool to release it all which made it pretty straightforward but upon opening the floor, the debris was far more interesting than I ever expected. I am not sure what I expected but I was surprised to see so much had dropped off the bees.

There was a multitude of pollen, lots of grey matter and then tons of what appeared to be rock salt. Having investigated this rock salt before I collected it all up claiming a miracle and a new hive product for us to all market, I realised these were the remnants of the wax the bees were secreting from their wax glands. Pretty cool, really.

Apparently tomorrow is going to be the hottest day of the year so far; glad I got that floor off in time.

JULY 10

D-Day. Two major events were happening today. Firstly it was the rematch between me and my traditional hive. They won last week and I had to retreat but, with my pride hurt, I've taken stock and today was the day. It was also the day that I was due to open up the Omlet Beehaus for the first time since the bees in the nucleus were moved inside. Two things I was interested in here. Firstly, had they settled in OK? And secondly, the nucleus contained standard frames and I had inserted these into the 14 x 12 brood box – standard frames are a lot smaller in depth. I wanted to see what was going on with the extra space they had at the bottom of the frames as apparently bees will build natural comb off the bottom of the frame to fill the void left there. I was nervously looking forward to it.

I decided to take on the Beehaus first as I know these were far calmer bees. When I took off the roof I noticed immediately that all the feed had gone and realised that the feeder was a little bit messy. Therefore I decided to take it off to clean it and then put in some more feed tomorrow for them. I then took off the extra supers, which you have to do when dealing with the Beehaus, and was amazed at what I saw. I had put five standard frames in and I could see immediately that not only had they built this beautiful natural comb on the bottom of these frames but there were bees covering at least eight frames. The natural comb was hanging down from the standard frames in a beautiful semicircle shape (they don't build comb in a perfect rectangle shape to fit the space but generally they build in this semicircle shape) and the hexagonal structure was perfect. It was amazing to see that in just a short time they had created this brilliant comb pattern. This was obviously good news as it meant that the bees were thriving and had been busy but to top that I also saw lots of lovely honey. It was all over the top of the frames and was a joy to see, especially when it was literally dropping out of the frame.

I have to say, it was all rather straightforward and it was good to deal with some nice, calm bees. It was great to see that the queen was indeed laying already into newly drawn comb and all seemed to be in order. It was also interesting to see the comb being built beneath the standard frames I had put into the brood box from the nucleus. It looked lovely and I knew that I hadn't had anything to do with it at all as I hadn't put my foundation down to guide them. I suppose this is what happens if you keep bees with what they call a top bar hive as you don't use foundation. These hives, rather than having frames and foundation like my hives do, simply have the top bar of the frame hence where the name comes from. The top frames only have a tiny strip of foundation or sometimes none at all and the bees then build natural comb from the bottom. I really like the philosophy of this more natural approach and really want to try it next year.

I closed up the hive feeling quite encouraged; just what I needed before I hit the rematch. It was probably a little bit like playing some weaker opposition just before a tournament to get a bit of confidence going – I was ready to take on the kamikaze pilots.

However, just to be on the safe side I filled up my smoker again, tucked my T-shirt into my extra-thick jeans – being great protection from the bees but really not a good idea on a 30-degree day – and then I went for it. I usually use the surgical gloves but I was pulling out the secret weapon, Jo's yellow Marigolds. It was almost a mission impossible as I needed to escape the kitchen cupboard without being noticed. How could I explain what I was about to do? But these would give me that little extra protection.

I was trying to come across as a new and modern beekeeper; I have a lovely khaki bee suit which looks as cool as a bee suit could get, and yet here I was with bright yellow Marigolds covering half of my arms. It's just as well that Jo was out of sight and no one could see me for miles.

My first job, once I had opened the hive, was to have a look at the frames in the super. Last week some of the cut comb was all over the place because of the heat which makes the wax very pliable. Usually frames of wax will have a wire strip going through them which gives extra support but these 'cut' comb frames are just open to the elements so I had to be very careful. When I opened it all up it was exactly the same and so I immediately replaced them back in the super and as they weren't drawn out, it was relatively easy. On reflection, not many of the frames were drawn out at all, which is a concern for my jar of honey. With the super frames replaced, it was time to move on and let the rematch begin.

Off came the super and it was immediately apparent there were a lot more bees even than last week. I gave it all a quick smoke and started to lift off the queen excluder. It was about this time that I again noticed, like last week, just how hot it was outside – I was baking inside the bee suit which is no surprise considering it was indeed turning into the hottest day of the year.

I started to work my way through and it took until the fifth frame for the kamikaze pilots to start. I have to say, it wasn't as bad as last week and I felt a lot more comfortable. The Marigolds were amazing and really helped to combat my nerves. It is strange seeing bees crawling all over your fingers but with these on I was a bit more at ease. I continued through the hive inspection and found that I was actually able to count the number of frames with brood which I had never been able to do before. Though nervous I was feeling far more in control.

I got to the last frame, having not found the queen but having found evidence that she was laying and so feeling quite satisfied.

I closed down the hive and as I left I was only followed by three bees this week. They stayed around for a little bit just to make sure that I was really going and then finally it was silent. I had made it; I had won. It felt great, not quite as relaxing as the Beehaus bees but I had got through it. Bit worrying about the lack of honey though. Have I missed the window of opportunity to get a jar of honey this year??

JULY 14

It started as a normal Wednesday evening session with the Reigate Beekeepers but it soon turned into all-out war with the bees. I learned some very valuable lessons! Beware of a windy night, especially if you are checking a colony without a queen.

I was with my regular group of six, checking over our hives, and it had started well. We were a little worried about the weather as it looked like the heavens would open at any moment. Still, we knew we had a good-tempered set of hives, though we couldn't account for the other twelve hives on the apiary. I was also told the weather, and in particular a colder and windier evening like tonight, can do funny things to even the calmest of hives.

We took off the top supers of the first hive and had got down into the first set of frames when the wind picked up out of nowhere. It was like a mini hurricane centred over our two hives. This, along with the slightly cooler temperature this evening, meant the bees were not happy.

We should have realised something was slightly amiss before we even started. Maggie, one of our mentors, had already been stung standing about fifty metres from the hive – I even saw the bee hit her fringe. The bee obviously decided she quite liked it and so buried itself ever deeper. Having got smacked out of the fringe during what I can only describe as an elaborate dancing move from Maggie, she went again but decided that the longer hair at the back was a better place to nestle. Despite Maggie's best moves, she got stung.

We should therefore have realised that some of the hives would be feisty but I am not sure we were ready for what was about to happen. Once we had taken out a few of the frames it really struck us that something was just not quite right. Firstly they were really, really angry – there were plenty of kamikaze bees flying into our veils – and secondly we actually lost count of the number of queen cells we had seen. This meant only one thing as we were certain we hadn't seen any eggs; the queen was no longer there and they were madly trying to raise a new queen.

This would account for the temperament issues but as we worked further through the hive, it turned into a war. Everyone kept on getting stung. Andrew took one on the finger then, almost immediately, Richard took one on the wrist. Then, as Andrew was checking some of the frames (we all took it in turns) he was stung again on the thumb but this time it must have hurt – I think I learned some new words! It was mid inspection, a frame of bees in Andrew's hand, and as he started an unrepeatable diatribe, the whole world went into slow motion when his hand that was stung instinctively left the frame. The entire frame of bees began to swing wildly and was being held by one hand. The hive tool flew through the air at the same time and landed about six feet away and you could almost see what was going to happen next... Fortunately Andrew gathered his composure before the frame dropped to the floor, spraying thousands of already angry bees everywhere. We quickly closed the hive up and for the second time in a couple weeks I cut an inspection short.

All was not lost, however, and we reduced the number of queen cells down to what we think are a few good ones. This meant that the colony may have a fighting chance of getting a half decent queen rather than taking a chance.

Fortunately, the other hive was a breeze but it was interesting having the kamikaze bees from the previous inspection still dive-bombing twenty minutes later. They obviously have a better memory than we all thought.

A pint was needed after this evening session, if only to look at all the war wounds and for Chris and I to gloat over our successful escape as we were the only ones in our group not to have been stung. Lady Luck was obviously with us tonight.

JULY 18

Now my bees are settled in the hive, it is essentially a waiting game to see what they can produce by the end of the season. I have noticed a distinctive shift in my attitude towards the bees. This journey all started with the setting of an aim to get a jar of honey and, if I am honest with myself, the caring aspect of looking after the bees came second. I knew they were in trouble but I was selfish in thinking the honey was what mattered.

I feel like I have changed. I have realised just how obsessive beekeeping can become and it is so obvious how fraught their situation is. When you consider that a third of our dinner plate is pollinated by the bee (though this is any sort of bee, including both honey and solitary bees; which as their name suggests live on their own but are prolific pollinators) you realise just how important they are to our survival. When you consider that a beehive of bees would visit several million flowers in one day and one person can only pollinate thirty trees a day it puts it all into perspective. Without bees our food options would dramatically decrease. This is not something I think any of us would like to see and as a result I just feel the need to talk to everyone about it and share my new-found knowledge.

On Saturday I took my parents up to the hives to show them what I had been working on for the last six months. I have to say, I think they were impressed but I couldn't really tell as I was just talking at them for about half an hour about how amazing bees are. That distant look in their eyes had appeared and so I felt it was time to stop. I sat them down behind the hives and got them to watch the bees flying in and out of the hive for ten minutes. Needless to say, they enjoyed the first ten seconds but I think that was it.

I did the same with Jo and Sebastian today. I could have stayed there all day watching the bees. I think Jo now has a quiet fascination with bees (given the level of knowledge she has gleaned from my endless ramblings, which I hear when she is recounting my new-found passion to friends and family). A twenty-one-month-old is not quite as enamoured by bee flight paths as a grown man, but he did pick some peas for the first time and seemed to enjoy them. One of my dreams was to have an allotment so that he could understand where food comes from. Lovely to witness him take the first steps of discovering the origins of food. I felt quite proud.

We also went to the Hampton Court Flower Show this morning and I found myself judging all the gardens by the number of bees in residence and making notes on any flowers to which bees took a particular interest. I left thinking I had lost my mind!

I got more concerned however, when I discovered I actually wanted to have a discussion about washing-up gloves. Having undertaken a successful stealth mission to take them out of the kitchen on Saturday and gotten away with it, I had decided that washing up gloves were the way forward but I wasn't convinced that yellow was the right colour, especially with the khaki bee suit.

Having posted a blog article about the best colour washing-up gloves for beekeeping, I then entered into a conversation about the matter. Who would have guessed it would be so topical? People said that bees love the colour blue and so you should avoid the blue and purple gloves as the kamikaze bees could start up again. The general consensus was that yellow was the best colour. Damn.

When I reflect on a day like today and realise that I have discussed rubber gloves and their effect on bees and judged a flower competition by the number of bees in residence, I have to reflect on how beekeeping has changed me. I would sincerely love to get a jar of honey this year but the overbearing feeling I have now is the well-being of the bees. I want to ensure they are well looked after and that everyone I know sees them for what they are; a minor miracle of an insect and one we should all be aware of and looking out for.

JULY 19

Will I actually make any honey?

Recently I have been starting to have some doubts and when I did the inspection on Saturday I was very aware of the lack of stores being built by the bees. If they are not building enough stores then there may not be enough for winter, let alone a jar for me.

The Beehaus bees were put in quite late and though they are now covering eight frames I am aware that they will want to cover the rest of the brood box before going up into the supers. I am very doubtful this is going to happen and this was confirmed by Adam when he saw a video of the hive. I have to say I wasn't expecting to get any from this hive so I am just going to concentrate on getting them through the winter and give them as much feed as possible.

The National hive stands most hope as the frames are fully drawn now but they have only just started on the supers. They have drawn out a small section on most frames with a little bit of honey deposited but this isn't nearly enough for the one jar of honey.

They have some great stores in the brood box which is good to see and positive for the winter but I am worried the honey flow has well and truly stopped.

JULY 24

I'm feeling a bit frustrated about the lack of activity in the hives last week but I know I shouldn't. I have also been conscious that all the books suggest you check the bees every seven days. Today is the seventh day but I didn't get a chance to check them. I don't suppose it is too much of a problem as we are out of swarming season but I still feel like I have a responsibility there to keep an eye on them and make sure they are all right.

I must be feeling guilty subconsciously as well because last night I started sleep-talking. This is something I have done since childhood but thankfully I no longer sleepwalk. Last night I excelled myself and even managed a full-blown conversation with Jo. The topic was bees and not just any old bees but swarming ones. Apparently I was advising Jo that we had to be careful to keep them away from the hedge. Once I had given my pearls of wisdom I apparently just turned over and went back to sleep.

Thinking ahead to tomorrow, it is a really important inspection in my mind. The weather has been good and so I am really hoping that the bees have had a good spell this week and have drawn out more comb and deposited some honey. It's funny, several times this week I have found myself thinking about the bees and wondering what they were up to, nervous about the next inspection and whether they have been successful for me. I cannot imagine ending this year saying 'and so I ended the year with nothing'. Not really the outcome I would want.

JULY 26

I used to be a big fan of *The Krypton Factor* when I was younger. I loved sitting down on a Monday evening to tune in to the unflappable Gordon Burns putting people through the rigours of assault courses and other tests that included a mental agility test. I thought these were evil until I worked for the Ministry of Defence during a university placement year. I was in charge of testing military recruits on their physical and mental performance in a range of environments. I once had to put some poor, spotty and smelly teenage army recruits into an ice bath and hold them down in the water to prevent them from jumping right out again. They had to then complete some manual dexterity tests while answering simple questions. Though the trials were cruel and the data analysis tedious and rather monotonous (it did put me off a career in research science!) it was fascinating and you could see the effects of environmental conditions on the brain's ability to coordinate activity. I used to think these were the hardest tests you could ask someone to complete until I started beekeeping.

I have found the tasks I have to complete far harder than any drill instructor or Gordon Burns-type would have me doing. Halfway through my hive inspections on Sunday, memories of *The Krypton Factor* came flooding back as I struggled once again with the job of counting. The basic and yet fundamental task of counting frames with bees on, and secondly frames with brood on, is becoming a thorn in my side at each inspection. It sounds so easy, doesn't it? And yet with the added pressure of all the bees, I believe it would be the best test for the Ministry of Defence military recruits to test their mental awareness under a stressful situation. I would fail miserably.

As usual I had completed the Omlet hive with no problem at all and it was a dream. I had counted eight frames with bees on and six containing eggs and brood, which was a good sign. I had even got a little bit cocky and started stroking them as they were so calm. This is exactly as it sounds and you just run your hand down the centre of the frame. I must stop thinking they are a pet Labrador! The only sticky moment was the point I realised that I was subconsciously holding my breath hoping that I would see the queen. As soon as I saw her I let out a large sigh of relief while saying 'Hello Queenie!', obviously delighted that I had seen her. The exhalation of air was right on the bees and caused a frantic five minutes of activity as they suddenly realised I was there. It was like an mass exodus from the frame and the only bee left was the queen.

After seeing the queen, however, I relaxed and the hive inspection was quickly wrapped up, safe in the knowledge that she was there and they all looked healthy. I got to the feisty hive and then it felt like I had Gordon in my ear whispering to me. They were a little bit feistier than last week but I still felt a lot calmer about it all, knowing I had my lovely yellow Marigolds on.

I was halfway through the inspection and realised that I hadn't been counting at all despite finding it easier with the Beehaus where I felt a lot more in control. With this hive I just seem to be concerned with not upsetting the bees and finding the eggs and queen, and my counting goes out of the equation. It is funny what happens to the brain while you are trying to reassure yourself that there is no way the kamikaze bees have a way in to your bee suit and that they won't sting your fingers and thumbs when you are picking up the frames.

Incidentally, while going through the National hive I received my first ever sting to the hand but my yellow Marigolds saved me. As I was removing one of the frames, I felt an almighty vibration and heard a great buzz – well above middle C – on my right thumb. It turns out that I must have accidentally squashed a bee while trying to pick up the frame. As a result I had the interesting view of seeing a bee trying to sting my thumb but it couldn't penetrate the glove. It was getting madder and madder though as it had done just enough to get stuck with its sting partway through the glove.

Generally a sting penetrates and the bee then tries to withdraw the sting, and in doing so rips off part of its abdomen; the sting then stays in the victim pumping in venom but the bee dies.

As I was watching, the bee finally pulled with enough vigour that she ripped away her abdomen and the stinger was left behind and, without wanting to be too graphic, she left parts of her insides with it. She struggled off, obviously satisfied she had dealt with the threat but I would think also quite astounded as she had lost the bottom part of her body. She flew off and I stood there watching this pulsating sting which was quite fascinating, especially as no venom was being pumped into my thumb this time. Despite the fate of that bee, it was an interesting sight to see and before I carried on with the inspection I removed the stinger and gently puffed my thumb with smoke to remove the pheromone smell, which would otherwise have got the hive angrier than it already was.

On this occasion, Gordon had got the better of me and the drill instructor was just laughing; it felt as if they were on my shoulders like little devils. Sadly I didn't get to see the queen – always frustrating – but I knew there were eggs so I wasn't overly concerned. I put the hive back together and wondered if the Ministry of Defence or *The Krypton Factor* had ever considered a test using bees. I went back home, had a cup of coffee and filled out my hive inspection cards with part skill and part guesswork. Still, always a lovely and fulfilling moment when you have finished a hive inspection.

JULY 27

I have been earning brownie points today and ended up at the mother-in-law's house to clear some of her garden. As all men know, brownie points make the world go round and whereas you get a point for each good deed, it is double points when you are talking about your mother-in-law. However, it was quite an upheaval to get to Wendy's house. Considering we were staying over, the amount of stuff we have to take for Sebastian is immense and it meant that Jo and I were exhausted.

Having spent most of the day clearing weeds and chopping back trees, not to mention the prickliest bush ever, I was taking a well-earned rest now that Sebastian was in bed and sitting down tucking into some ice cream and a can of coke (what a treat but such is the power of brownie points!). Jo brought out the post that had been delivered that morning and I saw there, on the top of the pile, was the new edition of *Beecraft*. Essentially this is the official journal of the British Beekeepers' Association and is actually quite an interesting read. I put it to one side while the vanilla ice cream and coke mixture took me to a new level of happiness.

Just as I was tucking into another mouthful, Jo leaned over, picked up the magazine and started to flick through it. I couldn't believe my eyes and was watching in awe thinking that maybe, just maybe my inane ramblings about the topic or the constant sleep-talking about bees was having an effect.

She stopped and was reading something intently for a moment, and then turned to me and said, 'Nice to see you fit the beekeeping stereotype so well.' Slightly bemused, I sacrificed my final mouthful, interested in what she may have been reading. It became apparent that she had read an article entitled 'What Makes a Great British Beekeeper' and this was essentially the findings from a recent study undertaken by the Department for Environment, Food and Rural Affairs (DEFRA).

It stated that the recent uptake of beekeeping has been from people who enjoy listening to Radio 4 (otherwise Radio 2 or

Classic FM) and the chances are they read a broadsheet newspaper at least once a week. All sarcasm aside, this pretty much describes me to a T and I was feeling pretty good about it. I would like to think it was portraying me as slightly highbrow and perhaps with a semblance of intelligence.

It was only then, reading further on, and seeing Jo giggle away that I read the part stating that the majority were over fifty years old (66 per cent in fact). Due to Wendy, my mother-in-law, being there, and not wanting to jeopardise any earlier points scored, I didn't rise to the bait and silently closed the magazine for later perusal. Jo just smiled, obviously feeling quite pleased with herself.

How rude!

JULY 28

I am a very rich green colour today as I'm full of envy.

All the signs are that the honey flow has stopped and I stand around with bated breath hoping that my little ones are still out there flying around trying to get the last of the nectar to make honey. During the last two weeks of hive inspections, I have been willing them to just make me enough for a simple jar of honey and have been practically begging them on bended knees to get a move on.

Then three separate beekeepers made comments that make me feel like my bees are the 100-metre runners trying to run a 400-metre race. They got out of the blocks fast enough but have been fading ever since!

It all started with an email from Adam, essentially my online mentor (despite the fact that we share a pint with other beekeepers each Wednesday, most of our correspondence is over email), who said his hives are roaring along in what appears to be the most productive year for a long time. In his own words, 'This honey flow is a once-in-a-decade event.' I wasn't put off by this, though maybe a little jealous; but also pleased that the bees had done well and this was exactly what they needed.

Then on Monday I received a taunting text from Richard who knew exactly what I was aiming for and it read simply: 'Just inspected hives, four sides of supers in one hive and three sides in other super with sealed honey, looks like will get a harvest this year.' Interestingly he got his nucleus about a week after mine so he must have a strain of bee that is more like a good 400-metre runner in the same race as my own and finishing with some vigour. A rather sarcastic yet jealous reply was swiftly sent.

As I was beginning to accept my fate, my level of envy increased a notch or two when I was contacted by a beekeeping acquaintance of mine on my Beginner Beekeepers Facebook page. Now this guy is like the oracle of beekeeping Down Under and he is as commercial a beekeeper as anyone. If Carling made beekeepers he would probably be the best beekeeper in Australia!

He started off by asking how I was getting on and we got into some pleasant chitchat about how everything was going. I mentioned in passing that I had put in quite a lot of work and was still hopeful that I might still get a single jar out this year. To this comment I could almost hear his laughter all the way from Australia.

He stated that he'd had a pretty good winter so far and his 3,000 hives – yes 3,000 – were pretty active. He reckoned he could extract about 100 kilograms (about 220 pounds) of honey, which I have to say I thought sounded a little low and so I questioned it. It was at this point that he corrected me and stated that he was expecting about 100 kilograms PER HIVE!! I couldn't believe it. Here I was talking to a guy about my struggles to get one jar from a hive and he was getting more than my own bodyweight in honey from each hive! I felt like a true whingeing Pom at this stage but I suppose this is the difference between a little lowly hobbyist beekeeper like me and a true professional, commercial beekeeper like him.

Sorry, did I mention that he said that some of these hives collected this honey in a little over four weeks?!

Truly amazing what these little creatures can do, isn't it? You just have to love them.

JULY 29

I had a fantastic beekeeping session this evening and felt like I learned a lot. Firstly I had to combine our two hives together to make just the one. The two hives haven't been particularly strong all year and there was a fear that separately they wouldn't survive the winter, but if they were combined they might.

We put the hives on top of each other with a couple of sheets of newspaper separating them. The idea is that the bees gradually eat through the paper and in doing so get used to each other's scent or rather pheromones. Combining the hives with nothing separating them would result in all-out war. I could just imagine a huge scrum in the middle of the hive until the last bee was standing. It didn't sound particularly viable. Who knew two sheets of newspaper could be so useful?

The next thing I learned tonight was how to use a flamethrower. What a way to feel full of testosterone! Basically you have to clean the equipment to make sure it is sterile and free of disease. The easiest way to do this is to burn all remnants of bees off the hive. I am not sure how this works with a plastic beehive like the Beehaus. That could have some interesting results.

This was one job I did not want to miss and so I elbowed my way to the front of the queue and got stuck right in. When I turned the gas on from the gas bottle there was that slight time delay and then the all-too-familiar hissing sound you usually hear when you switch on a gas oven. With a few butterflies in my stomach I flicked the lighter ready for my eyebrows to be singed at any moment. Hey presto! With immediate force the flamethrower was lit and despite my momentary jump backwards from the surprise at its strength I was ready to go.

The problem here is to remember that you actually have a lethal weapon in your hand. I had to stop myself singeing the ground and perhaps more worryingly, stop myself flaming the plastic pipe leading to the gas bottle which I managed on several occasions. Not a good idea. After a while you also get a little carried away and you have to remember to keep the gun away from your fingers as well. It sounds stupid but you are flaming quite detailed bits of kit and so I did find this a little bit of a challenge; a fun challenge though. All in all, it was not as easy a job as it looked and was quite dangerous. What a fabulous way to finish the session, though: must get me one of those next year!

JULY 30

It seems quite usual now to be stating that tomorrow's will be an important inspection. However, not only will I be praying, probably in vain, that the bees might have started to get me some honey, but there is a rumour of a serious bee disease locally and I sincerely hope my bees haven't got a problem.

When I did my beekeeping course I was surprised at the level of attention diseases were given; but I have since grown to understand that this is an integral part of the process of keeping bees. Knowing how to look out for, protect against and, if necessary, deal with diseases has become a very real part of my beekeeping life. In fact I would go so far as to say that disease is now my first concern when it comes to looking after bees. It really has become that serious.

Already this year I have done various varroa checks on my hives and fortunately I haven't seen any evidence of the mites. (If you have superb sight you can look for these mites on the bees themselves, and then you know there must be a pretty heavy infestation, but the alternative method is to put a board in beneath the brood and count the number of mites that fall through the hive in a given time period.) However, bees can suffer from 'brood' diseases, and two of the most well-known are American Foul Brood and European Foul Brood (AFB and EFB). Both of these diseases are taken very seriously by the Beekeeping Authorities as obviously they have very grave consequences.

When I started beekeeping I was advised to sign up to BeeBase, a website (it can be accessed at www.nationalbeeunit.com) which allows you to register your hive locations. This is important for many reasons but the most important is that they will be able to monitor disease problems locally and notify other beekeepers that may be affected. This is what happened this morning. I received an email from our local bee inspector informing me of an outbreak of European Foul Brood very nearby – in fact it is probably within three miles of my hives.

On this occasion, the exact location of the problem beehive is unknown as it is from an unregistered beekeeper who didn't want to give his details other than his general geographical area. This seems a little strange but it demonstrated to me the danger for us beekeepers if you are not registered. Apparently this discovery only came about by chance when the beekeeper was trying to sell a nucleus of bees and the person buying them noticed they were diseased and reported them. Sadly because they had met in a mutually convenient location, he wasn't sure of the origin of these bees. Bad, isn't it, the lengths some people will go to cover up their own mistakes?

I have therefore responded by email to the inspector to see if he wants to pop round to check my hives out as well. From my understanding EFB or AFB both need to be reported to the National Bee Unit who carry out work on behalf of the Department for the Environment, Food and Rural Affairs (DEFRA) so it is all pretty serious stuff. American Foul Brood is worse than the European as there is simply nothing you can do save burning your whole hive in a pit in the ground – imagine that. It is caused by bacteria getting into a cell and killing the larvae just after the cell is capped. The tell-tale sign for those of us checking is that the cell cappings will appear sunken and brown in colour rather than the usual white. The test is to agitate the cell capping with a matchstick and should it be infected with AFB, when you remove the matchstick, a residue rope will be hanging from the stick. Also, there is apparently a distinctive smell to AFB that people have said is a fishy aroma. European Foul Brood sounds like it is more treatable. It actually kills the larvae before the capping (essentially by competing for food and thus starving it) and the larvae will obviously have quite a different appearance than usual. There are two sorts of treatment: firstly by administering oxytetracycline (but only by an inspector), and secondly by replacing all the frames that the bees are on. This can be done by administering something called a 'shook swarm' – essentially achieved by shaking the bees on to new frames. If this fails and EFB remains then again you have to burn the whole lot.

I simply cannot imagine how a beekeeper would feel if he had to dig a hole in the ground and then set fire to the whole lot. That must be terrible.

Therefore, tomorrow I shall be looking more carefully than usual to make sure everything is OK. Apparently, because I have new frames and bees from a nucleus, I should be fine, but you can never be too careful.

JULY 31

This afternoon I approached the hives with rather more trepidation than usual. All appeared fine as I opened up and inspected the Beehaus. The bees are continuing to draw out comb and are building a healthy level of stores ready for winter. The bees are now covering about nine of the frames and have eggs on at least five of them so all seems to be cooking nicely in there. I spotted Queenie and she seems to be laying well and so I closed the hive up feeling pretty happy. There was no sign of EFB anywhere, which is a relief.

As I looked over to my feisty National hive while inspecting the Beehaus I could only wonder at the activity around the hive entrance. I can only describe it as thousands of bees gathering at the entrance clambering to either get in to or out of the hive. When I got round to inspecting it, the bees on the frames were relatively calm and yet there was mass hysteria outside the hive. The inspection went without incident, though I was disappointed not to see Queenie for a second week. However, I know she is there as there was again evidence of her laying recently. There were only two frames' worth of brood, which just shows that everything is starting to slow down a little bit. It's quite sad that autumn is fast approaching.

I saw that there was a little honey deposited into the frames of the super – then I realised it wasn't really honey, though, as it wasn't capped – this means that it was still nectar. It is my understanding that when nectar is deposited into cells it is 70 per cent water and this needs to reduce to around 20 per cent before the bees are satisfied that it is honey and cap the cell to cure it for another time. Sadly, none of these cells were capped meaning the chances of my getting any honey are becoming fewer and fewer. Unless I do something pretty drastic like move my hive to some late-flowering crop, it is unlikely that I will be getting any honey this year.

I have heard from discussions with other beekeepers that heather is particularly late-flowering and this has reminded me of Steve from the London Honey Company. He said he would move his hives to the heather later on in the year to get a late crop of honey but this was in north Wales somewhere. Apparently the really good areas of heather are in the north. As much as I have been storing up the few brownie points I have earned this year, I don't think it is conceivable to suggest to Jo and Sebastian that I needed the weekend to drive my beehive to Wales or anywhere else north of Watford. This meant the option of finding an area of heather is probably out of the question. I won't be put off, however, and so I wonder where I can find some heather locally, even a small patch, and whether I would be able to put my bees on there? Hmm...

So much to do and so little time and I am not hopeful of any success. It would be devastating to get this far and not manage my jar of honey but at least I would have learned some valuable lessons for next year.

AUGUST 7

Jo's friends were coming over for a 'girls' afternoon today which meant I was told to leave the house for a few hours. I started by feigning disgust that I was being asked to leave my own house while they all sat around drinking tea and gossiping. In the back of my mind however, naughty thoughts harking back to pre-fatherhood days were swirling around. These included:

- watching the football
- going to the cinema
- going for a walk
- going to a pub with an open fire, which I would just stare at
- meeting up with friends
- reading the paper over a leisurely coffee

And even...

going shopping

In the end I decided on taking shopping to a new level and went on the search for beekeeping equipment. Oh dear, what have I become? I think this must have been going through Jo's head when I told her. My poor wife. I'm no longer the husband she married.

I left the house with a spring in my step and made the forty-minute drive south to Paynes, the beekeeping store in Hassocks, Sussex. Driving down I was imagining it to be just a normal-sized, modern-looking shop on a high street but selling beekeeping equipment.

I was driving down a normal main road, which was looking suspiciously residential, when a sign appeared on my left-hand side – that was it – Paynes Bee Supplies. It led me down a tiny one-way private road, only just large enough for a single car to travel along safely, with houses on each side.

I came to a small row of houses and what appeared to be a large shed off to the back of it. And that was the shop. I couldn't hide my delight. A really well-known beekeeping company was basically being run from a shed!

I walked up to the shed (OK, it was actually quite a large shed with what looked like extensions on all sides, but did I mention that it was yellow?) and it all seemed deserted. I then noticed that its opening times, stuck on a board on the now closed shed door, were 9 a.m. to 1 p.m. at weekends. Damn, it was 2.30 p.m. I couldn't believe it, how stupid was I, not to have looked up the opening times before I left?

My luck must have been in however, as a man in his early forties came out of the house next door, said 'Hi, I'm Mike,' and promptly opened up the shed. 'I'm still here so I thought I would say hi.' I am sure I wouldn't get this sort of service at a normal shop on a high street one and a half hours after closing time.

My first impression when I entered the shed was one of delight. The first thing I wanted, a large green feeder like one I had borrowed earlier in the year from Suzy, was literally at my fingertips as I walked through the door. This made a nice change to my usual incompetent bumbling around supermarket aisles trying to find something.

The place was jam-packed, floor to ceiling with bee equipment. Everything you could conceivably think of was available; the sheer amount of equipment they had fitted into the shop was incredible. There were hive parts everywhere, not to mention frames, feeders, extraction equipment and even a variety of ratchet mechanisms. I could have stayed there for hours looking through it all and learning what equipment did what but I felt sorry for Mike who had opened up just for me so I felt it was time to go.

Then I became distracted by the wall of honey and I was rooted to the spot. I may now know that there are more varieties of honey than you can get from a supermarket, but this really took it to a new level. With the honey jars displayed together in a simple but striking manner it was stunning to see the different coloured honey next to each other. The collection of honey made up an entire wall spanning about 8 feet by 8 feet and it was like looking at a rainbow of honey colours with everything from runny honey to solid, more granular varieties.

It was literally covered with jars of honey in all shapes and sizes and from all corners of the globe – I noticed New Zealand,

Australia, Bulgaria and even Hungary – and it was a sight to behold. I didn't notice any Polish honey, however; those gents from the parking lot back in May this year obviously weren't that good at marketing their product outside of Lodz. I was pretty sure that Jo liked really runny honey and so I picked up what I could tell was the runniest honey, which was from Hungary. It was Echinacea honey and looked almost the consistency of water and beautiful with its rich golden colour. To be classified as Echinacea honey it must be a monofloral honey, which means that the hives that this honey came from have been placed in an area almost exclusively made up of Echinacea. That must be quite a sight.

Mike was lovely and really nice to talk to; apparently they have over 500 hives dotted around Sussex. Can you imagine? That could equate to 30 million bees at the height of summer! It takes over four weeks just to extract the honey, even though they have an extractor that fits forty frames at a time (the usual extractors fit only four or six). Amazing.

After about half an hour I decided to leave Mike to his Saturday afternoon. Having heard all about marking queens and knowing that I struggle to see my queen in the National hive, I took on some of Mike's advice and left with a 'queen marking cage', which allows you to isolate her without other bees around. While she is trapped you can apply some gentle pressure to expose her thorax and can then dab her with a special marker pen to leave a permanent mark. It allows you to identify her far more easily in the future. Mark advised me to go for a blue marker pen as he reminded me that each year the colour changes for your queen. In short, if you have a queen born in a certain year you have to mark her with a certain colour. This allows you to keep a record of how old she is but also, if they swarm, it allows the new recipient to work out her age as well.

I left a happy man but I arrived home to realise that I had mucked up. Jo actually liked really thick honey. Damn. Still, it was a really nice afternoon and I am really pleased I got down there.

AUGUST 8

Still no capped honey! This isn't going well and it's looking increasingly unlikely that I'm going to get my jar of honey this year.

AUGUST 14

I find it most perturbing at the moment. Everyone is starting to talk about the bees slowing down with their queens laying less, the honey flow being over and it feeling autumnal already. No, no, no, it should be the height of summer! I usually love August but this is getting me down.

It seems that as soon as I have got myself started on this whole bee journey it is already coming to a close with the season drawing to an end. My local association announced at the last evening meeting that inspections will start to decrease at the end of the month once all the extractions have taken place. Apparently after an extraction, beekeepers will generally check their hives every other week as the season starts to wind down.

To me it brings a whole new dimension to beekeeping and I will have to read up on looking after them through the winter. A lot of people feed bees and some even wrap the hives up to keep them nice and warm.

No honey for me this year, then. I must just get them through the winter. A shame really but there you go. It is funny, at the start of the year I was so desperate to gain a jar of honey but as the year has gone on, having now got my bees, I realise there is much more to it than just the honey. However much I would have loved to get a jar, in reality it is about helping them out. Now that I have these bees I feel a real sense of responsibility and wonderment towards them and I just want to make sure they will be all right going into next year. It's funny how your priorities change, isn't it?

AUGUST 16

I am the type of person that likes to bury their head in the sand and I feel I have been doing this recently. I know I should really be thinking about the bees' well-being as autumn approaches, but there is definitely an element of disappointment with the chances of a jar of honey diminishing. I just haven't felt like writing a lot recently. However, I have had a week off from work, where I went away with the family and I am back with a cunning plan (similar to those that Baldrick used to conjure up if you are a *Blackadder* fan like me – which perhaps doesn't bode too well).

There are many stories of heather honey floating around at the moment as it has been such a good year for honey production. This almost mythical honey is produced late in the season as heather usually flowers late in the summer once the nectar flow has finished elsewhere. The Yorkshire moors are renowned for having heather in abundance. One slight problem, however, is the fact that I am about five hours away from the north and though in one of Baldrick's madcap plans I would probably fly up there on the back of one of the bees, I cannot justify these geographical factors in my plan.

Therefore today I felt it was time to put on my thinking cap at lunch over a cup of coffee and a sandwich. My first approach was to put a shout out to my Twitter and Facebook followers to see if anyone knew of a sizeable area of heather that is local to me. This was an easy first step but I had no idea of where to go next. I was prepared to move my beehive somewhere in the south-east but no further. It left me with a pretty sizeable area to search and my resources were limited.

During these moments of pondering I received a good omen. Another regular at the coffee shop popped in and said hello. David, a chef who regularly walks in straight from work with the classic checkerboard chef's trousers, then said quite loudly, 'Ah ha, just the person – I have something for you!' Having absolutely no idea what he was talking about, I wondered exactly what he could have got for me. He disappeared back out of the cafe and wandered off down the street. Now David and I have rarely spoken but he had gleaned that I was a beekeeper, despite my appearance in a suit and tie, and that I was aiming for a jar of honey this year.

In fact it's become a bit of a regular question now on a Monday lunchtime. As soon as I arrive, Joe and Gareth, the two guys who run the shop, and David ask me how the hive inspection went at the weekend. Recently they have been met with a rather glum face and I was probably answering much like a teenager with a shrug of the shoulders and a grunt. It must have been obvious to them that I had lost a little bit of interest as I wasn't really in the mood to discuss the bees but I hadn't mentioned that I was not expecting the jar this year – more out of pride than anything else.

Three large slurps of coffee – and several crazy thoughts on how to get my hive to a heather area – later, David popped back with a blue plastic bag containing a small, unobtrusive parcel. He had been down to Cornwall on holiday, had seen this and thought of me. How very strange, I thought, as other than discussions about his outrageous gambling habit and my on-going beekeeping obsession, we hadn't really spoken. In fact, I had only worked out his name two weeks' previous.

Over another slurp of coffee I unwrapped the parcel, only to unveil a rather plain-looking box. When I opened it up, though, and dug through the tissue paper to get to what lay within, I was speechless. David had only gone and bought me a honeypot with a honey spoon! It had a rather quaint design but it had a very nice bee on the lid. I was really touched by this gesture and there and then I made a pledge to him and the cafe owners, Joe and Gareth, that I would now raise my game. Come what may I would get that jar of honey. I would then come into the cafe, buy them all a coffee and toast, demand Joe and Gareth to take a break and we would sit there together and enjoy the honey.

Spurred on, I have spent this evening planning. There hasn't been much success from my social media following. I was offered a few fields of heather up in north Yorkshire but I don't think that is feasible. However...

I have conjured up a plan to contact the National Trust to see if they could help. I have fond childhood memories of going for walks every Sunday with my father and our dogs but I am not sure whether it was the walks that I enjoyed or the warmth of the pubs we frequented when we had finished. Occasionally we would walk over a patch of National Trust land that was covered with heather. I have no memory of where this was exactly but it couldn't be too far away. I could have hit on a goldmine here; a light bulb had lit above my head.

I have tweeted the National Trust PR team, who have recently signed up to Twitter (@nationaltrust), to see if they can help and I will also call the regional office tomorrow to find out if they can put me in contact with the right people who may know where this land is. Given they are well-known supporters of the beekeeping movement, as I know some of their properties currently have beehives, I have my fingers crossed. Surely placing one of my beehives in an expanse of their heathland cannot be too much to ask, although I suppose an organisation like that would require lots of health and safety regulations. That could be a headache. Let's hope not.

AUGUST 17

As I logged on this morning there it was, a tweet back from the National Trust: 'Hi @surreybeekeeper, we would love to help you if we can', and they then proceeded to email me the number for their regional office and the name of the person to speak to.

I phoned the office and having got passed around a couple of times I ended up speaking to the head gardener of Polesden Lacey, a most beautiful property not far away, with gardens designed by Capability Brown. I knew that they had just put a hive in the gardens there and we started out talking about that. He sounded fascinated by the project and was obviously enjoying having the hive in the gardens. I took my chance and explained my predicament and it was obvious from the outset he had some sympathy for my cause and stated he was keen to help if he could. However, he did not have any control over that particular area of the National Trust. He thought it sounded like Frensham Ponds, which rings a bell with me and is in the area that Dad and I used to walk – I must check this with him later.

He offered to email the head warden for that area, which I was really thankful for and, as luck would have it, he is apparently a very keen beekeeper. I also managed to get his phone number. I couldn't believe my luck; this was looking better and better. I just hope he will allow me to put my hive on their land.

Perhaps, just perhaps, there may be a way I can get this elusive jar of honey.

AUGUST 18

'What a Difference a Day Makes' was the first dance at my wedding and it is rather apt here too. I have gone from literally being unbelievably down in the dumps to being excited once more. This is all to do with the prospect of my one jar of honey.

Having waited twelve hours for the head warden to email me, I decided to take matters into my own hands and call him up as I wasn't sure how much time I had left. The last thing I wanted was to get the OK, to move my hive and then to realise that the heather had finished flowering.

I got through immediately and introduced myself. Fortunately he hadn't been ignoring me; he just hadn't checked his emails for the last twenty-four hours. What a way to live your life; I have become far too dependent on them and it takes someone like this to jolt you out of your little world.

I started to give him the quick précis of what was going on as I had done yesterday but I reckoned I had now got my pitch down to a T, keen but not too desperate, and tried to big up the National Trust for helping me. I even got to the point of saying he couldn't refuse on the principle of helping a fellow beekeeper – OK, perhaps that did sound a little bit desperate.

Once I had got through my five-minute diatribe on why I needed help, he simply responded, 'What a fabulous idea! I am sure we can help.' I LOVE THE NATIONAL TRUST!!! He then proceeded to tell me not to think about the original area I had considered for various reasons but another area, which I was familiar with but not aware of its heather credentials; the Devil's Punch Bowl near Haslemere. The reason he knew this to be such a good area was that he lived there, and just a short walk away from his home were several hectares of flowering heather. My heart was beating far faster than usual and I felt on top of the world.

Until this moment I had kissed the jar of honey goodbye and now the door was back open again to hopefully get something out of the year. I had genuinely never thought I would get this reaction from the National Trust and had feared it being a little bureaucratic in its outlook. Here I was however, with not only an answer but a positive answer at that and all within a ten-minute phone call.

We talked further about bee-related things and then got down to the subject about what would be flowering at the moment. Apparently the heather isn't always a great harvest but thrives in wet summers – hence the last two years have been pretty good. This did mean that this year wasn't looking great as it has been so dry but the last two weeks of wet weather may have helped me out. He did state that the heather had been flowering for a little while and he only hoped there would be enough nectar left for the bees to utilise.

He pointed out another flower which I should look out for, which is apparently going great guns at the moment, rosebay willow herb – I have never heard of it but if it helps then I promise to plant it in my own garden next year.

We talked about heather honey for a bit. I knew it wasn't like any other honey but I hadn't really considered that you need to extract it differently. Usually, when you extract honey you have to spin it around at high speeds and it all shoots out of the frames easily. Heather honey is quite different and it simply won't do this. It will stay in the frames stubbornly refusing to move. Removing heather honey is a lot more labour intensive and involves pressing and squeezing it out of the frames. Great, just my luck, I thought – but then again I only need a jar of it.

The head warden then started talking about a mate of his, however, who had designed a contraption that was made out of an old tumbledryer drum and spun at some ridiculous speed; enough to defeat the reluctance of the heather honey and throw it out of the frames. I suggested that this guy sounded like some sort of engineer. 'You could say that,' was the reply. 'He designed the engine for Concorde!' I nearly dropped the phone. I have to meet up with this guy, I thought. What a fantastic way to extract my honey under the direction of a Concorde designer. I made it a personal goal to meet this guy if I was in time for the heather nectar flow. This would be a far more entertaining way of extracting the honey than manually squeezing it out of the frames.

Having taken all of this in, it just left a few moments to get things arranged and work out exactly where the head warden lived. Looking at the map, I thought I lived in the middle of nowhere but this guy took that to a whole new level.

Saturday it is then. I will be moving my hive like a commercial beekeeper in America. I will literally be chasing the honey. Morally I know this isn't great as travel puts stress on the bees but I am only talking about one hive and a forty-five-minute drive. This is hardly the 3,000-mile journey the American bees make on the back of huge lorries, which usually takes two or three days. No wonder they get stressed out.

AUGUST 20

My preparations began today and I wanted to take a look in the National hive and assess the state of play before the move so that I could plan ahead. I was also interested to see whether there was any more honey in the super, just in case I was being a bit premature in this honey chasing.

I lit the smoker as usual and took off the hive roof. There seemed to be a lot more bees than usual up in the super which I took to be a good sign. Taking out the best super I could see that only a little bit more honey had been collected since my last inspection. However, it was really nice to see that about five or six cells had been capped over, which means that the consistency of the honey was right. Five or six cells, though, is hardly enough, considering the other frames didn't really have anything on them at all. This just confirmed that I am doing the right thing.

Stupidly, I was rushing a bit today and so was putting the frames back into the super quite quickly. This soon backfired as one frame, ironically the better frame with some capped honey, fell out of my hands. It was similar to watching toast fall from your hands at the breakfast table, spinning around in slow motion on its way down to the floor where it annoyingly lands butter side down. The same happened here but this time there were quite a few bees on it as well. I couldn't believe it. What had I done?

On picking it up gently, I realised there was no damage done except a few displaced bees who were a bit perplexed, as moments ago they were on a lovely yellow smooth frame and now they were fighting with long green stems of grass. I replaced the frame a little bit more carefully this time, relieved that I hadn't lost what little honey had been collected.

I lifted off the super and started to look through the brood box but no sooner had I done this than I felt a sensation like a mosquito bite on my leg. I couldn't believe it, especially as I was midway through lifting out a frame of bees. I calmly put the bees back down into the hive and then as a precaution I swatted at my leg. Walking away from the hive, it dawned on me that it was highly likely to have been a bee sting, especially as I was standing on the ground exactly where I had dropped the frame earlier.

As I lifted up my trouser leg there was indeed a little sting hanging out of my leg, just above the ankle. In fact it wasn't just one sting, there were two little stings pulsing poisonous fluid into me. I was somewhat perplexed as it really didn't feel like a sting. However, the tell-tale sign was the long yellow stain going down my leg – this must have been the poor bee that got squashed as I swatted at what I thought to be a mosquito.

I was no longer tucking my trousers into socks so this time the bees must have crawled up underneath my trouser leg to attack. There seems to be no option but to look into buying a full-on boiler suit next year. They just seem to like my lower limbs and there is no way of guarding against these ankle biters.

I got back to the inspection, but decided only to check the National hive today as I was a little pushed for time and I was only really concerned about planning for the move. I was really pleased, though, as I saw the queen for the first time in weeks, but stupidly I didn't have my marking equipment with me. (I say this like I would have thought about carrying this out; there is no way alive I would have been able to mark her on my own.)

I went back home, had a cup of coffee and concocted a plan for tomorrow morning. I think I have decided on the following:

1. Go up really late this evening and put a gauze over the entrance to stop the bees flying out.
2. Tie a ratchet around the hive to enable me to move it all in one piece.
3. In the morning lift it into the car and drive to the new site.

It all looks so easy when you write it down over a coffee; but having done a move before I now feel a lot more at ease with it all.

AUGUST 20 – EVENING

OK, the move is underway. I first of all needed to test whether the hive would fit in the car as a complete entity, as there would be a problem otherwise! It was already obvious that the roof of the hive would have to come off but with the cover board in place, the bees would be safely inside. Before I walked to the allotment I took a tape measure and measured the height of the boot and then once I was standing behind the hive, I measured that as well. There were literally millimetres in it – too close to risk doing it early tomorrow morning and failing to get the hive into the car. I decided I needed to do a dummy run on getting the hive into the car just in case.

I went back to the house, drove the car up to the allotment and reversed rather gingerly until I was about 10 feet from the hive. I tried to secure gauze around the hive entrance before securing the hive using the ratchet. I didn't fancy lifting up a whole hive of bees for them to come pouring out to see what was going on; that wouldn't have been pretty. My first attempt didn't go particularly well and the hive wasn't secure; the bees found it very easy to navigate around my drawing pin-secured gauze. They must have just thought I was a complete fool. I left them alone for a moment hoping that they would lose interest as I went to the Man Shed to find some gaffer tape.

I entered the messy oasis of the Man Shed, but none could be found and my hope for electrical tape was also in vain, so I had to put up with layering the rather feeble masking tape. I wasn't too concerned as it was only a temporary measure tonight to just move the hive into the car, but I would need something a little more robust for the actual journey.

Back at the hive, the bees had indeed popped back inside and so I secured the gauze with the layered masking tape; it did look a bit of a mess but I wasn't overly concerned. I started using the ratchet to secure the hive together and all was going well, having watched someone using one on YouTube earlier for good measure. They are amazing little bits of engineering and my hat is docked to whoever designed them.

The ratchet was secured and I started to lift the hive. I did everything right and bent my knees, not my back, but none of this was going to help. OH MY GOD. As I strained to lift the hive, I could not believe how heavy it had become since I moved it there nearly three months ago. I could hardly move it. I had no option, though, as there was no one else around and so I had to have another go; I only needed to lift it about ten feet to a car boot.

I lifted and stumbled my way to the car, all the while listening, rather too closely, to the bemusement in the hive. I rested it on the lip of the boot and as I levered it forward a little, it dawned on me that I was correct earlier on. It was very, very close to fitting in and there were millimetres in it but not in the right direction. To be honest, now that I looked at it very closely, the hive should actually have fitted in given its measurements but it was the angle in which it had to go that was causing the problems, so I didn't feel too much of an idiot. There is a small lip on my boot that I had to ease the hive over first before sliding it in and this was causing the problem.

It wasn't the end of the world. I just had to remove the hive stand on the bottom, which would give me another 10 centimetres to work with. Off I stumbled back to the hive area, and off came the ratchet to let me prise it all apart.

This was where it got tricky as I tried to break the stand away from the hive. Even using the hive tool it was practically impossible. I couldn't get any purchase on lifting up the hive except with the stand. Because the super and brood box were filled with bees I couldn't be too heavy-handed, because they weren't fixed together. The last thing I needed was to force it and to dislodge the super from the brood box, and thus let countless bees out of the hive that probably wouldn't be too happy.

It must have taken about half an hour of nervous and scary manoeuvres to create a small gap between the bottom of the hive and the stand. As I was fixing the ratchet I was literally putting my ear right up against the hive and all I could hear was the noise of 50,000 bees hemmed in, trying to work out exactly what was going on. I tried not to think what might happen if I dropped the hive off the stand. Just as I managed to get the ratchet in place, the stand wobbled and the hive moved about one inch off the hive stand, onto its dodgy and rather weakened leg – the one I had put the misplaced nail into all those months before. I saw it wobble, knowing this one leg couldn't take the weight, but fortunately managed to grab it in time. I had a very, very tense moment or two. The sound of the bees was terrifying, as it must have felt to them like an *Italian Job* moment, when the bus is precariously placed over the side of a mountain pass. I started to breathe normally again as I realised all was OK.

I lifted the hive to the car once more but this time without the stand, and fortunately it fitted in fine. A wave of relief came over me as I realised the move was still on for the morning so I put the hive back on the stand. Having taken off my bee suit I put it and the rest of the kit I would need into the car and I was pretty much ready for the morning. The only thing I would have to do tomorrow morning is bulk up my entrance block. I am sure there must be something on the market that is far more professional than a gauze/masking tape bodge job but it's far too late in the day for that. I just need something that will keep them in place for forty-five minutes before I unleash them on their new home, which I am still to see. I wonder what it is like.

My last job this evening was to simply open up the entrance again to allow some ventilation. This was pretty stupid as I had put everything in the car ready for tomorrow, including the suit. I couldn't be bothered to put it all back on again and as it was getting dark I didn't think there would be a problem. After all, it sounded quite calm in there now. I bent down beside the hive and prised the masking tape off and then gently took out the pins which held the gauze in place. I then gently took away the gauze and all was fine. For some reason I then decided to shine the torch at the entrance and WHAM, two guard bees came straight at me buzzing frantically. Why oh why did I do that?!

I ran as fast as I could from the hive, leapfrogged my raised beds and headed straight for the exit of my allotment. They were still after me and weren't letting up. I continued at a rather fast walking pace towards home and after about a hundred metres it went quiet, much to my relief. However, all was not over and they must have regrouped, plotted and planned. Five metres further on they made another sustained attack and this time they must have split up. As one attacked my head, the other went in for the kill on my now exposed belly seeing as my arms were waving frantically around.

One sting on the belly button and then within about five seconds I got a second. Ouch, they really hurt this time but I suppose that will teach me a lesson. As I trudged back in silence I made a mental note never to shine a torch at a beehive entrance again.

All that aside, I am hoping that it is like a heather heaven when I open them up tomorrow morning at their new home as they fly out of the hive into a river of nectar.

AUGUST 21

Having had four stings in twenty-four hours, I was feeling a little bit silly today. I mean, how desperate must I be to get this jar of honey? I wouldn't have got stung had I not wanted to move the hive. There must be a moral there, or maybe the bees just know something is up.

Anyway, when I got back last night I searched high and low for an alternative to masking tape and gauze and found... nothing. No electrical tape or gaffer tape as I had hoped, and no foam to block up the entrance either. I was stuck with the method I had used last night, which I didn't have a lot of faith in. However, I decided to take it one stage further and so I was now armed with masking tape (lots of it), the gauze and then tea towels; yes, that's right, tea towels.

I was going for a thick masking tape primary layer, gauze secondary layer (just in case they nibbled through the tape during the journey) and a tea towel third layer. I thought it an unbeatable entrance and would tide me over until next year when I would get something far more professional.

It was 5.30 a.m. and I was there early with the plan that I could get them into the car quickly, drive to the location, set them up and be back in time for breakfast. As I approached the hive and got it all secured it looked like it was a good possibility. My plan was coming together and my three-layered entrance block looked great and was very secure.

I set off feeling confident and it wasn't long before I was driving in thick woodland close to the tiny turn-off I needed to take. I was heading down a main road towards the Devil's Punch Bowl with my eyes peeled looking for a sign advertising a youth hostel. The only problem I had this morning was that the head warden wasn't home and so if I got lost I was on my own.

Ahead of me I saw it and immediately slowed down. There was a tiny unmade road off to the right-hand side. It was just wide enough for a car, with banks up both sides which prevented me from seeing over them. As I went further down the track and the banks gave way to views of woodland in all directions, it immediately felt remote. I realised it must have been quite damp because I kept seeing mushrooms and the woodland floor was covered with a thick, almost fluorescent green moss. The mushrooms were not just any mushrooms mind you; they were those beautiful ones that look like toadstools and will kill you in an instant should you take a bite.

The head warden had mentioned that it was a little bumpy down the track. You could say that again. As I continued down, his words were ringing in my ears. I had thought I would be fine as I knew our farm track was pretty bad, and I had therefore taken his words with a pinch of salt. I was very glad I'd come in the four-by-four, as at points it felt like I was about to tip over. My main concern was the bees but they seemed fine.

The tiny road kept winding round and round but was dropping in height quite considerably as I went down into the most beautiful valley. It was like a hidden world. I went past a very basic-looking youth hostel where, according to my father, he stayed many years previous, and carried on. The road narrowed further and it all became rather atmospheric with the trees feeling like they were smothering you; everything was almost too green.

I turned a corner and dipped down once again, and up on my right I could make out the most beautiful old house covered in moss, the gardens overgrown in a nice and natural way, and the whole thing looked like something from a fairy tale. This must have been the head warden's house. I stopped the car and looked up at it for a while. It was one of the most unusual houses I have ever seen and given my history as an estate agent, there have been many.

My instructions told me to drive past the house and then park in a little turning bay up ahead, which I found pretty easily and stopped the car. Behind me, over the boundaries of the fence to his garden, were the remnants of some beautiful WBC hives. Obviously they had fallen into disrepair but they looked lovely there, and so fitting of the setting. It answered one question for me: at least my bees wouldn't be fighting with others.

As I stepped out of the car, it was immediately apparent how peaceful it was. Nothing moved around here. It was all still. I did a 360-degree turn taking it all in and that is when I saw the heather heaven that I was looking for. At the end of the road, there was a cattle grid and the valley opened out. In front of me were the rolling hills of the Punch Bowl and all I could see was heather, all in flower, which made these rolling hills look like soft pink blankets. I couldn't help but smile. I only hoped I wasn't too late.

I walked over to the cattle grid and into the valley for a few minutes, taking a closer look at the heather. I had never considered it much before now, but heather on this scale is really quite stunning and I couldn't wait to get the beehive out of the car to let them at it. It struck me again as I excitedly walked back to the car just how quiet it was out here. It was as if the air was still and time had stopped. The heather just stretched for miles.

I found my way through the heather and into a more wooded area, locating the point that the head warden had told me about. What I hadn't really considered in my plan was finding an appropriately flat piece of land. It all seemed so uneven and rough, not exactly perfect for a beehive which you should try to keep level at all times. I hadn't thought to bring a spade to level an area off, or even a paving slab to give some semblance of a sturdy base.

After some searching I found a small area which seemed the best of a bad lot. It was out of the way, so that no walkers to the area would come into direct contact with it, and the hive stand would go either side of an old tyre track, now covered with grass. It was pretty much level, and so I went about my duties.

With the hive now on the stand I was ready to let them out. I positioned the entrance of the hive so that it was facing the heather so that the bees could fly straight out to this glorious sight. All looked good, and learning from yesterday's events, I opened up the hive with my bee suit on. Almost immediately a few bees came out to see what was going on. It was lovely to see a few fly out straight away, presumably to see where they were.

I made a quick retreat back to the car, conscious that I needed to get a shift on for breakfast, but I couldn't resist a quick admiring look back at the hive. It looked perfect there, with the white of the hive against the pink of the heather, and I got back in the car a happy man; I only hope that I am not too late. There is not much I can do but sit and wait.

AUGUST 25

I really want to go and check on the beehive today and see if the bees have settled in to their new home, but I know I shouldn't. I wonder what it was like for the bees to suddenly have all those flowers on their doorstep – it must have been magnificent. I imagine it was like a disco in the hive as all the foragers came back in from their travels and performed their respective waggle dances to inform others of where the good sources of nectar were.

The waggle dance is an amazing phenomenon which in the darkness of a hive allows bees to communicate with other workers through means of vibration. A worker will come in and perform a figure of eight movement around other bees. The angle at which they perform this manoeuvre represents the direction that the particular flower can be found, and the intensity of the waggling, which they do as they draw the middle section of the eight, implies exactly how far away it is. Apparently it is accurate to within a foot in a 3-mile radius, which I find pretty mind-boggling.

I have to refrain from going to the hive for two reasons. Firstly, I think it is common practice to leave them to get on with things for a couple of weeks to get settled; and secondly, and perhaps more pertinent, I am off on holiday for a week.

I did manage to call the bee inspector, however, who sounded far nicer than I expected, following on from his email which was rather abrupt and to the point. I explained my hive move and that I only had one hive but he said it was no problem and was more than happy to come over anyway and check the single hive. He said it was just as important to meet up with beekeepers as it was to see their hives, which I thought sounded like good enough logic. I was confident that I didn't have a problem and from my descriptions of my hives, he was too. He again alluded to the fact that I was using 'fresh comb' and this is usually a good thing. We made an appointment for early September.

SEPTEMBER 2

I have just come back from a mad weekend with Dad and the morris dancers. It has been a tradition for the last forty-five years that my father has returned to the same campsite on August bank holiday after he got lost trying to find Wales with his mates. They made it as far as a pub called The Kings Head in Withington, and he and his morris men have been going back ever since to commemorate the occasion with the same landlady, who is now in her eighties. Amazing, hey?

I have come to the conclusion that beekeepers are now a little cooler than morris men and in fact I have decided that beekeepers are not as fat as morris men nor are they as drunk – shame on them – but I always have an entertaining weekend with them, where we stay up far too late, sleep in tents and generally have a wonderful time.

As I returned back to normality I realised that tomorrow is my 'bee inspection'. Oh Christ.

Despite the bee inspector sounding nice on the phone I still have images of this rather traditional, stickler-for-the-rules beekeeper coming round to my house to inspect the bees. I can see myself now, standing to attention beside my hive: fully suited up, veil on, hive tool ready to go and smoker lit as he inspects my outfit first. Something similar to the way that Prince Charles inspects the troops while on parade; one hand behind his back while the other pinches a bit of fluff off the uniform. Engaging in polite chitchat, the bee inspector will then make small talk in a rather formal, very English manner. 'So then,' he will say, 'you think you are a beekeeper do you?' Oh Christ, I will be thinking. 'What is your opinion of the new Bailey Comb Change method?' I would faint rather than offer an answer.

On a more serious note, I will be interested in what he thinks of the Beehaus. I wonder if he has inspected one of these before. I am sure he must have done but I am also not really sure what he will be looking for. It will be nice to see if he thinks I am on the right track, and not fundamentally screwing everything up. He arrives at 10 a.m. tomorrow. Must go and prepare everything and make sure it is all in order (tea and biscuits a must, I would say). Fingers crossed.

SEPTEMBER 3

Of all the stupid things I could have done, I have stood up the bee inspector. I feel like an athlete who didn't turn up for a drugs test. I forgot about Jo having a midwife appointment today, and usually I go to these for moral support. I have had to put an emergency call in to the bee inspector. I could only leave a message. Huge black marks for me but there are times when other things are of far greater importance.

So far Jo is getting on really well with this second pregnancy and, for the second time, we are not finding out whether it's a boy or a girl so it is all rather exciting. We both feel it is life's greatest surprise and I remember so clearly the anticipation with Sebastian as he popped out and, as we had requested, I got to tell Jo that it was a little boy. I couldn't think of any other way I would like to deal with the situation again and I simply cannot wait to become a father again.

On another note, I am going to look at the hive on Sunday. Hopefully it hasn't been knocked over or vandalised while it has been there. In the meantime, I had better eat some humble pie and apologise to the bee inspector when he returns my call.

SEPTEMBER 5

Today was the day of the first hive inspection in the heather fields. Had the move worked? The weather had been good for the last two weeks and so I was hopeful. I also decided to make it a bit of a family outing and so took everyone along. Even Mum and Dad wanted to come along as it wasn't far from them, and secretly I think Dad wanted to reminisce about his younger days staying in that youth hostel.

We drove down and met my parents by the beautiful house of the head warden who was still away on holiday. It was so weird seeing other people in this little haven of peace but what was stranger as I looked out to the heather was that there was now a herd of cows walking around out there in the valley. They certainly weren't there last week.

As Jo, Sebastian and Mum and Dad walked around and introduced Sebastian to the cows, I got down to business and opened up the hive with great anticipation. Going through the well-rehearsed routine of lighting the smoker, then suiting up, followed by puffing the smoker, I was getting quite nervous of what I might find. It was going to go one of two ways; either a full super of honey or nothing would have happened at all.

As I looked through the hive, it was perfectly clear that it was to be the latter. Aside from a few more cells in the super being capped, nothing seemed to have changed. I was gutted. The fact that the hive was particularly feisty did not help matters. The only difference I noted was that some of the outer frames in the brood box had evidently been filled with honey as they had definitely changed and got a lot heavier. It was also interesting to see that the honey was far darker and thicker than I was used to. Maybe now these were filled they might do something in the super. I could only hope. I would leave it here for another week at least and see how they got on.

As I finished off I noticed that not only were Mum, Dad, Jo and Sebastian standing some distance away watching, but also some walkers had stopped to see what was going on. To top it all off, a little group of cows was now standing about fifty feet away having a nose. It seems beekeeping is of universal interest.

I walked back to the car with a straggler hanging around me but by the time I got there it had disappeared. I disrobed and started packing away. Whilst talking to Mum and Dad, however, it appeared that the straggler who had been guiding me away from the hive came back. This time she seemed to take issue with Mum, and particularly her hair. Just as with Maggie before, it started on her fringe before making its way to the longer hair at the back. However, unlike Maggie's rather elaborate dancing manoeuvres, Mum was somewhat more controlled which was surprising. She went for a sustained swatting technique with almost robotic arm movements and the occasional yelp of 'Get that bee out of my hair!'

Fortunately, this time there was no sting and when we managed to get the bee out, she must have thought her job was done and promptly flew off. Feeling a little flustered, Mum made a quick retreat into the car and we duly followed. It was time to go home. Deep down I was a little disappointed but it had been a great experience going out there and I will be interested to see what happens in the next week.

Forgot to mention: despite the message I left yesterday for the bee inspector, it seems that he actually turned up to meet me. He called to say that he got my message too late and had already arrived at our house. That is surely going to count against me when we do finally get a chance to meet up. He agreed to pop back in a couple of weeks but I feel pretty bad about making him come all the way out to see me again.

SEPTEMBER 9

I spoke to a few of the beekeepers at the association today and explained my predicament. It seems that almost everybody has now extracted their honey and is enjoying the fruits of their labour. Many have been saying that it has been the best flow of honey for years. They all said that getting a jar or two in that first year is always a difficult affair, but starting as late as I did would make it practically impossible. This just confirmed what I feared all that time ago but was probably not willing to admit.

A few were understandably sceptical of my motivations of trying to get a jar of honey and thought I was mad to go running around the country on the search. However, there were others who could see what I was trying to do and some suggested that I take a little bit from the brood stores. After all, if it was just enough honey for one small jar then it shouldn't affect the bees too much, especially if I planned to feed them as well. I had never considered this before but it gave me a viable backup plan. I wasn't particularly keen, but they had plenty of stores in the hive based on the inspection at the weekend so maybe, just maybe.

Originally I was planning for a nice large jar of honey to put on the breakfast table but now I'm thinking about altering this plan. If there is not enough honey in the super and I take some from the brood box, I don't really want to take too much after all. I was given two of the smallest glass jars by Omlet as some of the freebies that came with the Beehaus. These weren't quite on a par with those tiny plastic jam or honey capsules you get in a hotel but they weren't an awful lot bigger. As they are glass, I am sure I could get away with calling these a jar. Perfect: my standards have lowered to an achievable aim and I wouldn't be affecting the bees too much.

SEPTEMBER 12

I made the decision to leave the hive at the Devil's Punch Bowl for a little while longer as I would like to give them the opportunity to take on as much of the heather as possible, so I had only the Beehaus to check today.

I didn't see the queen but I did see that she had been laying recently, which was encouraging. There was one thing within the inspection that truly astounded me however; the 'waggle dance'. Now I have been doing plenty of inspections since I started back in June but I had yet to see a waggle dance at full throttle. It was fascinating. I thought originally that they just did a waggle and then disappeared to get on with other duties. Seeing a bee doing a waggle dance in a particular area and then moving across the frame a little to do it elsewhere, to then move again was really interesting to see. It probably explains how information can travel so fast through the hive. These magical little messengers were simply mesmerising to look at as they spread their good news.

When I was watching this amazing dance I was struck by the thought that I hadn't used a lot of smoke on this inspection. Maybe I had been smoking them too much previously and it was probably like dancing through the smog-filled backstreets of London during the 1950s. They must have given up, thinking that the darkness is bad enough, let alone this smoke stuff.

Regardless, it was lovely to see this dance for the first time in my own hive and not on a video of someone else's. Yet again I was reminded of how amazing they are.

SEPTEMBER 13

Something quite unexpected happened today. I have mentioned before that I live on a farm track. The farm at the bottom is owned by a gentleman by the name of Steve, a quite elusive character who keeps himself to himself. In fact, other than wave at him as he drives by our house, I have never really set eyes on him.

From what I have heard from others, he isn't the nicest of characters and certainly he doesn't get on with Farmer Ray who lives opposite. Let's just say there is a lot of historical politics going on which has got quite unpleasant from what I can understand. So I was quite wary of meeting Steve, who by profession is actually a carpenter and had now turned his hand to farming. Imagine my surprise, therefore, when his car pulled up next to me as I was walking down the track.

His big black van with tinted windows drew to a halt and the windows rolled down to expose a slightly gaunt face with aviator-style sunglasses. He must have been in his fifties and it looked like he hadn't had the easiest of lives. I am not sure how I can describe the greeting but it was certainly abrupt, and the entire conversation went like this:

Steve: 'You're the beekeeper fella, aren't you?'

Me: 'Yep, that's me, I just started this year and have a couple of hives, one is actually elsewh–' He cut me off.

Steve: 'Good, I have several thousand apple, pear and cherry trees up there at the farm. If you want to put your hives there feel free. I asked the beekeeping association before but no one returned my call.'

Me: 'Wow, that sounds amazing, yes, thank you. I am moving my hive back this weekend actually. Would it be OK to mo–' He cut me off again.

Steve: 'The code is 1897 for the lock, make sure you lock it up behind you, I am having a problem with gypsies at the moment.'

Before I could even say thank you and goodbye the window was being rolled up and he was accelerating at speed away from my standing position in a cloud of dust. I couldn't really work out what had just happened. Not only did he appear to confirm what most people had said about him in that he was rather curt and to the point, but he had also offered me an opportunity which was just unreal.

Most beekeepers dream of a site like this, wherever it may be, let alone just up the road. I couldn't believe my luck but, if truth be told, I also couldn't believe that I had never seen this orchard previously; it sounded huge. I had to get up there before I went to collect the hive.

SEPTEMBER 18

I am trying to do lots of reading about how to get the bees ready for winter which I understand to be a combination of disease prevention and feeding. Aside from lots of finger crossing from the beekeeper, the bees need stores which they can call upon in the darkest depths of winter until the spring comes along once more. Apparently one of the most nerve-wracking moments for beekeepers is during the warmer days in spring when they look for activity around the hives. This will tell them whether their hives have survived the winter as you very rarely check them when the weather is cold.

There's not much else I can do at the moment, but it's going to start getting busy over the next few days as I prepare to extract some honey, whether from the super or the brood box, and move the hive back home again. However, this time I am moving them back to paradise. An orchard! What a lovely surprise that will be for them as they fly out the next morning... Actually, that is probably a lie as they are simply trees at the moment with no blossom. It's nice to think, though, that they will fly out, look at the trees, instantly recognise what they are and decide that this winter will be worth sticking out because spring will be a fantastic time with lots of blossom to be had.

Before going to get the hive tomorrow I wanted to go and have a look at where I was going to put the hive. I have never been near Steve's actual house before as it is rather imposing and the two great big Rottweilers outside, looking like they would eat my arm off should I even look at them, never gave me a good enough reason to go and say hello.

Therefore I got in the car, Sebastian by my side (thinking along the lines that if a man wouldn't hit another wearing glasses, dogs wouldn't eat an 'intruder' with a child) to make the small trip up to Steve's farm. I arrived and it was immediately apparent as I helped Sebastian out of the car that no one was there, not even the Rottweilers. The huge house, complete with now run-down outbuildings, was deserted in a rather eerie way. It must have been a huge operation at one point and from what I hear from Farmer Ray, it used to be a dairy farm; this may be where some of the confrontation lay, as it used to be Farmer Ray's home where he grew up. Now these outbuildings were all in terrible disrepair with fallen-in roofs, some of them covered in ivy. There were sounds similar to those you hear in Westerns or when you see disused airports in a desert and random mechanical sounds reverberate in the background. It really felt quite uncomfortable.

I crept up to what I thought was their front door and peered through the glass windows to see if I could see any signs of life. WOOF, WOOF, WOOF!! The two Rottweilers leaped up at the door with snarling teeth, snapping in front of me. As I jumped out of my skin, Sebastian, on top of my shoulders at this point, simply went, 'Hello doggy, nice to see you doggy.'

Having quickly left without so much as a 'Goodbye doggies', Sebastian and I went for a walk through the orchard. Most of it was behind the house, hence why I had never seen it before, but some of it touched the road where a huge embankment had blocked our view previously. It was like heaven, rows upon rows of beautiful fruit trees, probably planted three or four years ago looking at their size. There must have been at least a thousand of them and Sebastian and I just kept walking around looking at this wondrous sight. I cannot imagine how magnificent this must look in spring with all the blossom: it must be a sight to behold.

As we continued walking we came across two slightly smaller orchards, again filled with fruit trees, evidently planted at a similar time. It was just beautiful and I left satisfied, though a little bit worried about how this was going to pan out. I didn't know where Steve wanted me to put this hive therefore I rang him on his mobile and found he was in the pub. He said to pop up a little later, at about 8 p.m., and he would show me where he thought it would work best.

This worked out fine for me and so Jo and I got Sebastian all tucked up in bed before I popped out in the car once more. This time I tried his gate to gain access to the orchards from the other direction, which is ultimately what I will be doing tomorrow rather than going in via the house; I don't fancy carrying the hive all that long way. It unlocked first time, and on this beautiful September evening I made my way up to his house though the fields, the twilight of the dusk now surrounding me. Halfway there, down a dirt track, I saw Steve in front of me, with his eldest son by his side who was probably no more than nine years old. There he was, shotgun cocked and in his finest hunting regalia, and as I pulled up next to him he looked even more daunting than usual.

I lowered my window, gave him a friendly hello and was met with three grunts as he pointed at three different areas of his vast orchard. Within a couple of minutes he walked off in the opposite direction and I drove off, feeling rather excited, in the general direction of his pointing and had a quick look.

The first area looked a little too close to his house and so I discounted that immediately. If I was to expand the number of hives next year it wouldn't be entirely practical to have them too near the house and, if I am honest, I am a little bit worried about those dogs because if they got close it wouldn't be great for them. Their aggression would be no match for several thousand bees. The second option he had suggested was an area towards the back of the orchard which looked simply beautiful. On one side there was open fields and to the other was the orchard. I could see that in the springtime it would be stunning, but I also felt that it could be a little exposed. The third option was by far the best. Protected by a hedge and yet with a direct view into the orchard it looked fantastic. There was enough room for expansion and also it was far enough away from the house.

I got out of the car and stood where I feel my hive will go tomorrow. It just feels right and I cannot wait to see what it will look like next year with the blossom out. I cannot believe how lucky I am to have this on my doorstep.

Thinking ahead to tomorrow, I have only one slight fear; the thought of being stung again while moving the hive. I must remember to take it easy and not to rush. I just can't deal with the itching all over again.

SEPTEMBER 19 – SUNDAY

Another D-Day has arrived. I have enlisted the help of a good friend of mine, Jeff, who I recently heard had wanted to become a morris dancer therefore making me think he would be perfect for the job of going to get the beehive. Bearing in mind last time I struggled to lift the hive there was no way I was going to do it on my own this time.

There is another ploy here as well. Jeff and I met when Jo and his wife, Kate, were both pregnant and attending these slightly unorthodox childbirth courses. While the adults on the course bonded over the hilarity of our course tutor's love of herbal medicine and its effect on childbirth, our children have also grown very close since. Sebastian and their little one William, now nearly two, have become inseparable.

However, Jeff and Kate were brave enough to have gone for the second one a lot earlier and Luke popped out in June this year. Therefore my little journey with Jeff today is also to ask him a whole load of questions about how on earth he is coping. Bearing in mind he has bags under his eyes which give the impression he has been out shopping for the whole of the country, I am not looking forward to the feedback I might receive. Even so, I need to know what having two children is like and this is the perfect opportunity.

We left early on a slightly autumnal-feeling morning. I can't believe it is coming around so quickly again. It was absolutely beautiful at that time of the morning, however, and there wasn't a soul around. Hopefully it would be early and cold enough that the bees wouldn't have started flying around yet.

We found the turning and got down to the hive pretty quickly. Looking around it was evident that the cows had been moved on again, just as my bees were about to. In the distance I could make out my hive and all looked quiet.

I had mentioned to Jeff that I only had one suit and was happy to give it to him while moving but the first job was to secure them in the hive once more. I pulled out my mess of a contraption to do just this. It looked like one jumbled heap of masking tape and tea towel but it had seemed to do the job. I approached the hive and all was silent. I quickly attached the entrance block, securing the mass down with drawing pins once more. It was all over in a matter of seconds, which was perfect.

I gave Jeff my suit and the move went pretty smoothly. I had left the ratchet mechanism in place so we just lifted it all straight into the car. The good news was that the hive seemed that little bit heavier than when I had lifted it into position previously. Perhaps there was hope. Within five minutes of getting there we were driving back out again and I was feeling rather pleased with myself as there was not a sting in sight. Result.

We arrived back at the house in good time and rolled up outside Steve's gate. In no time we were standing in the same place I had been not even twelve hours earlier. Jeff and I first had to move a honking great piece of limestone into position to provide a solid base for the hive to stand on. I wanted to give this hive a good foundation. Who knows if I would ever move it again, and so I wanted to give it the right start.

With Jeff and I both puffing loudly, we managed to manoeuvre the limestone into position and the stage was set; it was time for the hive. We positioned the hive stand and then the hive on top. It looked wonderful. There is nothing nicer, in my humble opinion, than seeing a beehive in an orchard. I wasn't sure about bringing my freezer box of a hive in here as it might not look as good, but this hive was spot on. I couldn't help but smile.

It just left me to open them up and welcome them to their new home. I was excited, perhaps overly so and because I had given Jeff the bee suit, I opened up the entrance block and just ran as fast as I could back to the car without looking back. It was perhaps a tad dramatic but I didn't want to take any chances given the way the bees had reacted last time.

I knew that very shortly I would be back up to take a look inside. In fact, thinking about it I will probably leave them a bit and open it up for the first time in a few days when the bee inspector comes around. Here I am, about four months into my career as a beekeeper and I am about to be inspected: scary stuff.

SEPTEMBER 22

The inspector, Alan, arrived. I very carefully peered out of the window as he walked up the path. I couldn't believe it. There I was yesterday scouring the web for a photo of an American drill sergeant to add to the text of my blog and here was one walking down the path toward the house. He had the heavy boots on and the hat was an exact match. That was where the comparisons stopped, but I had a good giggle at the hat.

As I opened the door, he was immediately friendly and helpful, putting me at ease as he explained what was about to happen. Alan wanted to crack on and so we headed straight up to the Beehaus as it was closest. I didn't really know what to expect from this inspection but he immediately got stuck in looking around for any varroa – there was barely anything, which is a relief – and all the signs from the hive were good. He did suggest I remove a couple of frames from the hive as they hadn't been drawn out yet. This would allow the bees to concentrate on building up stores rather than wasting energy building up frames. In a flash it was over; this inspection seemed a lot more straightforward than I thought.

We went to my National hive in its new position. Not having seen it since I ran away late on Sunday night, I was quite eager to see it again. Fortunately it looked quite good in its new location and all looked well... Until we looked inside.

The last time I fully checked the hive was September 5 and all looked well. Going through the hive today it was immediately apparent that there was a queen cell a few frames in, but it was now empty! This could mean only one thing: that there was a new queen somewhere, and according to Alan it was most likely a supersedure, where the bees decide they want a new queen but, having produced one, don't swarm. It was quite likely here as the number of bees in the hive was not noticeably different, and these events usually occur later on in the season.

We carried on and I was lucky enough to spot the queen, but she was definitely not the same queen. She looked a little bit smaller than Cleopatra – apparently characteristic of a virgin or new queen, and so I felt quite pleased (and reassured) that I had seen her.

There were at least another couple of queen cells in the hive as well, but all had been opened, meaning other queens had emerged or the bees had destroyed the cells themselves. This would have meant there would have been an almighty queen battle in the hive. Queens are only able to sting other queens and so this is what would have happened until the best queen won and only one remained. Alternatively the bees could have been happy with the queen that had emerged and decided to rid the hive of the other queen cells. Therefore something has certainly been going on since September 5! Typical that it was just in time for the bee inspector to see it all happen as I feel that it has made me look a little bit inexperienced. It may be that I missed one of these cells in their early stages before I moved the hives back from the heather. Queen cells take twenty days from the moment the bees decide they want to raise a new queen to when she hatches out. Alan was very nice about it and did say that in a busy hive they are easy to miss, even for experienced beekeepers.

All in all I received some great advice from Alan and it was well worth the visit as it was actually really good fun and I feel that I have learned an awful lot. Aside from Richard who was simply filming me, Alan is the only other person who has been there with me while I have checked the hive and it was nice to have the company, even if I did feel a little daunted at times. Alan did mention that it is time to start feeding the bees which I must get prepared for and buy the right sugar this time. This will ensure they have enough stores to call upon over winter and Alan believes that the Beehaus needs a lot for the bees to survive. On the other hand however, he says that the National has plenty of stores available, obviously a good sign!

SEPTEMBER 24

A rather apt day, being my son's second birthday, I will forever remember it being the day I extracted my first honey.

It was the moment I had feared. I'd heard horror stories of beekeepers chancing their luck removing honey from the bees late on in the season with the general mood in the hive not exactly being excellent. Bearing in mind my feisty lot, it could be fun. Having taken advice from a few beekeepers recently, I had decided to take the honey from a frame in the brood box. It was evident from the inspection with Alan that there was nothing in the super and so desperate times called for desperate measures.

I tried to think of the human equivalent of what I was about to do to the bees. If you can, imagine spending the whole summer stocking up on groceries and stores for the winter; baking cakes and making stews to make sure there are plenty of supplies while the winter weather sets in. Then you spy your slightly irritating neighbour, who insists on coming in once a week just for a chat, walking down the path to your house. You sit there patiently, knowing what is going to happen, only for him to open the back door and signal for all the other neighbours in the street that the path was clear. People then walk straight into your house and steal all of your home-made produce and stored fruit and veg for the winter. I am not sure you would be happy and you would probably come out fighting, aiming to wallop any greedy neighbour with a broom.

There could be an upside, though, as they may just leave you a roomful of sweets just to make sure you don't starve. Once I'd removed the frame I would be putting on a great big feeder complete with sugar solution which they should be able to refuel with once I replaced the frame. As Alan had said, this was a pretty strong hive with good levels of stores and so a little bit of burglary wouldn't hurt them.

Once we had put Sebastian down for a lunchtime nap, I went up to the hive, filled with trepidation. I went through the motions: smoked them, left them for a little bit, then opened it all up with the hive tool. I started choosing the frame and it was as if they knew what was about to happen. They were worse tempered than usual. I just needed to go through the other frames, choose the best one and then get out. I quickly selected a frame, shook off the bees, brushing those still hanging on with a bee brush (a very soft bristled brush), and made a swift exit.

I drove home and promptly spread a small tree's worth of paper around the kitchen and got down to business. Having realised I was not going to be taking several frames of honey I didn't bother hiring an extractor and so I thought I would go about it in a rather manual way. Therefore this morning, after Sebastian had opened his presents I had got all my equipment prepared.

Usually you would use a serrated knife to decap the cells but this wasn't going to get the honey out of the cells. I therefore decided that I was going to scrape the honey out of the cells using my hive tool. I needed somewhere where the honey would spill into but I didn't really have anything suitable. I ended up finding a bright pink cat litter tray which was perfect for the job and so having cleaned it out and put it through the dishwasher a couple of times, it now had pride of place in the kitchen. Of all the pieces of equipment in the world that were at my disposal I never expected to be using a bright pink cat litter tray for honey extraction but there you go – welcome to my world!

Having set it all up earlier, with the hive tool in one hand and the frame of honey in the other, I thought I was ready, like a well-oiled machine...

... That was until I saw a honeybee flying around my kitchen. I couldn't believe it. I had arrived not even two minutes earlier and already they had found me. I checked that all the windows and doors were closed as I know that bees will follow you, but I was at least 500 metres from the hive, I had driven the long way back with the frame in the boot and then walked straight into the house. Yet this little bee had found me already. Amazing, but surely a coincidence? I let her out the door, hoping that she hadn't sensed the honey. I had to move quickly just in case before she told all her mates.

I gathered up my tools again and rested the frame on top of a piece of kindling wood that was straddling the bright pink cat litter tray. This was to keep it clear of the honey while it dripped out, which was a little tip I had seen online. I gently pressed the hive tool into the frame of honey, being careful not to push the hive tool through the complete frame. As soon as I increased the pressure and it pierced the first few cells, honey oozed out over the rest of the still-covered cells. It was beautiful to see and I couldn't resist running my finger over the golden liquid to catch some before it dripped into the tray. This was to be the first time I tasted my own honey and was an experience I will forever savour. Jo, who had been standing behind me, held out her finger to catch a droplet or two and it really felt special as we both stood there enjoying the moment. As I continued to press the hive tool in carefully, I was struck by how soft the honeycomb was and how little effort I had to put in to extract the honey. It was a real joy but it was over too soon. I only had one frame to do and within a few minutes it was finished. I was looking at a frame now empty of honey as it was all in a waxy mess in the bottom of a bright pink cat litter tray. I was not sure any other beekeeper had ever extracted honey the way that I had just done but I felt pretty pleased. I wrapped the frame up and put it to one side. I would deal with that later.

For now, I would let the honey settle and then push it through a muslin, probably tomorrow. In the meantime, I had birthday duties to attend to.

SEPTEMBER 25

It's amazing what you can get done in a two-hour naptime once Sebastian goes down for a sleep. Jo and I rushed downstairs into the kitchen and immediately put a muslin over a ceramic bowl that we had sterilised that morning in the dishwasher. You use the muslin to take out most of the impurities, the wax cappings, for example. As Jo held the bowl in place I lifted up the litter tray to let the honey flow out onto the muslin. For the first time, as the honey was filtering through the muslin, I really got to see the colour of the honey, as the bright pink of the litter tray didn't really do it justice.

It was lovely and golden, and as I expected, it seemed to be the colour and consistency of that urban honey I had seen previously. It was however a little bit thicker than I imagined and hence I had a feeling this task might take a while.

After a few minutes it was pretty clogged up and so we lifted the muslin out of the bowl to take a look at the underside. None had dripped through and I feared I had used too fine a muslin but, as we were looking, a large drip dropped through, which was a most joyful sight. After that came another, but it was clear it was not going to be a quick job.

Jo and I therefore decided to leave the muslin in place for a while to let it all filter through. Securing the muslin in place using some books surrounding the bowl, we left it and let gravity do its work.

Coming back to it several hours later once Sebastian was in bed, we saw that most of the honey had now filtered through but there was still some stuck in amongst the wax cappings that were left there. I picked up the muslin, held it over the bowl and then did what came most naturally at the time; squeezed it as much as I could until I was satisfied there was no more honey that could come through.

Once this was done we could just go straight ahead and bottle up the honey. Looking in the bowl there was definitely enough for one small jar and we may have been lucky enough, in our ultra-crude way of working, to secure two small jars. I could hardly contain my excitement. As I held a jar, Jo very carefully started to pour the honey. It was the moment I had been waiting for and as the first stream of honey dribbled into the jar it all felt worth it. The jar slowly filled up and the honey was a beautiful golden colour but filled with miniscule air bubbles. If I had done it all properly, I should have left the honey to settle for a period of time, which causes the air bubbles to rise and escape, but I didn't really have the patience this time around, I was too desperate to try a little bit of it!

We quickly switched to the second jar which, with a bit of help from a spoon to encourage the last few droplets inside, meant we had two full jars of honey. I was chuffed to bits. Here in front of me were two jars of my own honey. I had to take some photos for posterity. I had succeeded.

Jo popped down the toaster and within two minutes out came two slices of toasted white farmhouse bread. Wasting no time to sit down, we spread over lashings of beautiful butter, watched it just melt into the toast and then, using a knife, we harvested the last dregs of honey from the bowl. As we spread it over the toast we got that wonderful waft of honey and Jo was straight in to her slice. Since she is not one to mince her words, it was lovely to hear her say how nice it tasted in just one word: 'Stunning'. I took an almighty mouthful myself, and it was immediately apparent how strong the flavour was. This wasn't a mild honey by any stretch of the imagination and was, honestly, one of the nicest honeys I have ever tasted. The fact that I have only tasted about five different varieties doesn't really matter. This was my honey – what a lovely thing to be able to say.

Jo and I just sat there smiling, accidentally putting another slice of bread in the toaster. It was a wonderful moment to share and while we waited in anticipation for the next slice I used a teaspoon and filled up the honeypot that David had given me, and prepared myself to fulfil my promise on Monday at the cafe.

SEPTEMBER 26

When I started this journey into beekeeping I viewed it as an education for both Jo and I but also for Sebastian. I want him, and hopefully baby number two, to grow up and understand where food comes from and hence why I am passionate about growing my own fruit and vegetables as well. Beekeeping seemed a natural step for me in this quest as well as helping out bees at the same time.

I therefore had to sit with Sebastian on this relaxing Sunday morning, and take him through the same experience that Jo and I had last night. As he sat there in his high chair looking up at me I felt immensely proud of being a father to such a wonderful little boy.

In front of us was some fresh toast, some butter and my small jar of honey. I quickly explained to Sebastian what was going on and then started to spread the butter over the toast. As it gently melted I quickly stuck my knife into the jar of honey and started spreading.

The fusion of melting butter and honey looked delicious and I cut the toast into small strips and offered one to Sebastian while Jo and I took the other two strips. As I put the toast into my mouth the fusion of honey and butter was immediately apparent. The warmth of the toast combined with the coolness of the butter still melting and then the depth of flavour to the honey was simply delicious. I looked at Jo who seemed to be enjoying her toast just as much and then my eyes settled on Sebastian.

His eyes were like saucers as he looked at the scene around him. He would look from Jo to me and then down at the strip of toast in front of him, not really sure what to do. After thirty seconds or so he picked up the toast between his thumb and forefinger as only a two-year-old can manage. As the toast curled and covered his whole entire hand, he went in for the mouthful.

Though never managing to bite into the toast given its angle, Sebastian proceeded to lick off the entire honey/butter concoction, before throwing the rest of the toast back down on the breakfast table. With his chin covered in honey and butter he looked at both Jo and I and simply said 'More' or rather 'Mowa' in his slight baby-like tone.

Jo and I just looked at each other and smiled. All the hard work to get that single jar had just paid off and I was a very happy man. Only the honey-tasting in the cafe to go now...

SEPTEMBER 27

Looking back now, had I not gone to the cafe that day, had my usual discussions with Joe and Gareth about the bees and then seen David with his gift of the honeypot, I don't think I would be in this position I am in now. Therefore it seems apt to end this story on the day that I sat with them and fulfilled my promise.

On reflection, looking back over the year, it really was the turning point. I'd been pretty disappointed that I had put all this hard work into beekeeping and I wouldn't get a jar at the end of it. Joe, Gareth and David had all got me going again.

It had taken nine months to come together and finally here I was. A jar of honey was looking back at me, a lovely and enticing golden yellow in a small jar. I had stuck my finger in previously and knew it was lovely but I hadn't shared it around.

I approached the cafe, honeypot in hand, and entered. The place was deserted save David in his usual seat at the back, and Joe and Gareth standing to his right catching up on the day's news. I approached, demanded toast to be put in the toaster and popped the honeypot in front of David, who instantly recognised his gift from a few weeks' previous. With a sinister smile he suggested that I could hardly call this a jar of honey and then proceeded to break out in a huge beam as he shook my hand. It was a great feeling as we all sat there discussing and laughing at whether this could be classified as a jar of honey. They truly knew, perhaps more than most, how hard I had worked to get it. The next ten minutes were a joy as we sat there undisturbed, Joe and Gareth taking the highly unusual step of having a break and sitting down at the tables I had seen them serve at and clean for the last three years. Unusually, no customers came in and we enjoyed each other's company and tucked into honey on toast. Needless to say, it was delicious. It was bliss, four people coming together over a honeypot and a cup of tea.

It was soon time for me to get back to the office and so I thanked them all and left. I was sad somehow that it had to come to an end. Closing the doors behind me, it really seemed like it was the end of the year. Both hives were now being fed to get them ready for winter and the beekeeping season was drawing to a close. I had started the year as someone with a passion for gardening and the outdoors and with a growing interest in beekeeping and here I was, someone now utterly obsessed by the subject.

I have made many mistakes this year, much to my annoyance. I have learned that perhaps I am not as organised as I like to think I am and that I can sometimes take a too relaxed perspective on things. But all in all I have learned one very important lesson. If you have a passion and believe in something enough, you will always succeed. Where there's a will there is a way.

It was the end of a fantastic first year's journey. I walked away from the cafe knowing I'd had an amazing year, learned the most mind-boggling new hobby which I am keen to enhance and improve on next year, and have a very exciting winter to look forward to. While my bees are all tucked up warm in their hives this winter, I will become a father for the second time. Life doesn't really get much better.

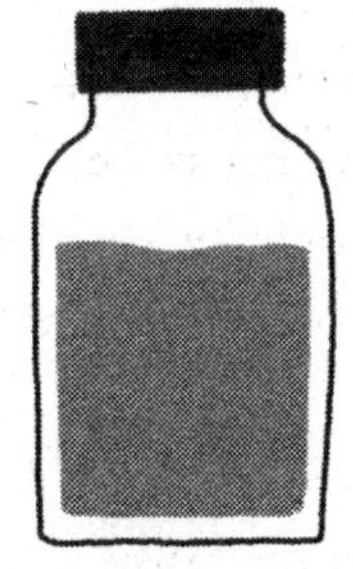

USEFUL RESOURCES

Publications

BBKA News
www.bbka.org.uk
A free publication that you receive if you join up with the British Beekeeping Association, which I highly recommend you do when starting out.

BeeCraft
www.bee-craft.com
Britain's best-selling beekeeping magazine.

Bee Culture
www.beeculture.com
The USA's best-selling beekeeping magazine.

Hive Lights
www.honeycouncil.ca/index.php/hivelights_home
A magazine for Canadian beekeepers and a good read (worldwide subscriptions are available).

The American Bee Journal
www.americanbeejournal.com
Specialist publication available by subscription including beekeeping information, education, classes and events, as well as beekeeping history.

The Australian Beekeeper
www.theabk.com.au
A very comprehensive magazine for all Australian beekeepers (worldwide subscriptions are available).

The Beekeeper
http://nba.org.nz/publications
New Zealand's premier beekeeping magazine (worldwide subscriptions are available).

The Beekeepers Quarterly
www.beedata.com/bbq.htm

Websites

www.abfnet.org
The American Beekeeping Association.

www.bbka.org.uk/help/links/other_associations
This is the website for the British Beekeepers' Association (BBKA), which is great for finding out course information or details about

local beekeeping associations. This page specifically leads you to information about other national associations such as the Scottish and Welsh Beekeepers.

www.beekeepingforum.co.uk
Probably the most popular beekeeping forum based in the UK, with international topics regularly discussed.

www.biobees.com
If you are interested in 'natural beekeeping', this popular forum is for you.

www.honeyo.com/org-International.shtml
This site has an extensive list of all bee-related associations you could conceivably want. It provides specific information on beekeeping in the UK, Canada and the USA too.

Facebook

www.facebook.com/adoptabeehive
A worthwhile campaign run by the BBKA, and just in case beekeeping isn't quite your thing but you still want to be involved in combating the bees' plight.

www.facebook.com/beecraftmagazine
The page for *BeeCraft* magazine, Britain's best-selling beekeeping magazine.

www.facebook.com/beginnerbeekeepers
A great online resource for all beekeepers.

www.facebook.com/britishbeekeepersassociation
The page for the British Beekeepers' Association.

www.facebook.com/savebees
A great community page for everyone that loves bees and beekeeping.

www.facebook.com/solitary.bees
A useful page for those who are interested in solitary bees rather than honeybees.

Blogs

www.surreybeekeeper.co.uk/my-recommended-beekeeping-blogs
My own blog, so I hope that you find it useful. This particular page contains a list of blogs that I constantly update, so keep checking back.

www.adventuresinbeeland.wordpress.com
An entertaining read by Emily Heath.

www.alananna.co.uk
Alan and Anna's stories about their simple life in west Wales – about beekeeping and self-sufficiency.

www.beekeeping-book.com/blog
Off the back of David's excellent beekeeping book is his blog, containing some fantastic photos.

www.danieljmarsh.wordpress.com
Though a new beekeeper, Daniel has jumped in with both feet!

www.drypulse.blog.com
Follow Joseph's blog as he starts beekeeping.

www.geommm-bees.posterous.com
George, a long-term Twitter friend, keeps this great blog.

www.gibbshoney.com/blog
Russell is a fourth-generation beekeeper, and his blog is well worth a read as he develops his hives from two to eight.

www.kiwitopbarhive.blogspot.co.nz
Marcia's excellent blog about keeping bees the natural way in New Zealand.

www.likerockpools.wordpress.com
In his words a 'humble' blog, but I love it. Well worth a read.

www.lilsuburbanhomestead.wordpress.com
An informative blog about beekeeping and all matters to do with sustainability.

www.maconhoney.blogspot.com
A well-structured blog written by Macon, with some great photos.

www.marks-bees.blogspot.co.uk
A nice blog from Mark, who is based North Carolina, USA. He has been keeping bees for four years.

www.nosolomiel.blogspot.com
Jandro's Spanish beekeeping blog.

www.reenysbutterfliesbloomsbees.blogspot.com
Maureen's blog about adapting her garden in Florida, USA, for butterflies and bees after her original garden was destroyed by Hurricane Andrew twenty years ago.

www.romancingthebee.com
Deborah's blog about beekeeping and gardening, with a real sway for 'making urban beekeeping beautiful'.

www.snoodlesdoodles.blogspot.com
A simple blog with some great images, which covers a range of sustainability topics, beekeeping being just one of them.

www.stevensbees.blogspot.com
A very comprehensive and visually interesting blog by Steven, who is based in Dudley, USA.

www.suburbanbeekeeper.com
An informative and useful blog by Will.

Twitter Contacts

It would take me forever to list the full website addresses for these Tweeters, so I have listed their Twitter handles. All of these guys are worth following and communicating with, and have been great fun to talk to online:

@surreybeekeeper (OK, so this is me but I hope you follow me!)

@AFBR
@AlisonBenjamin1
@AnneWareham
@annoyingserf
@ApiMaye
@ascorbic
@BeeBeebytheSea
@BeechwoodBees
@BeeFriendlyZone
@beekeeping
@beesfordev
@BeesinArt
@BeesInFrance
@beesknees42
@beeware55
@britishbee
@BurtsBeesUK
@camlad_apiaries
@cornwallhoney
@cotswoldbees
@DamianGrounds
@DavesBees
@DawnIsaac
@DCHoneybees
@EdenCaterers
@egglatina
@ElevensesTime
@Emily_Heath
@EmmaSTennant
@EvansBeehives
@farmingfriends
@forkmagazine
@GardenerGareth
@GeoMmm
@HDoodles
@HelenReeley
@helpthebees
@HenCorner
@hillbillytilley
@HuwSayer
@IanDouglas
@IBRA_Bee
@insidebooks
@jake_schultz
@JaneStruthers
@Jimmmy_Bee
@KarinAlton
@Kate_Bradbury
@KiwiManaBuzz
@Lancasterbees
@LincolnGreen
@LisaCoxGardens
@Loiscarter
@LondonBeeKeeper
@londonsbuzzing
@lottieplot21a
@MaldonBeekeeper
@maradadisimba
@MartinGBEdwards
@McKellier
@mimisbees
@MISSSWhitehouse
@mizzlizwhizz

@nigelsbees
@oh2bnMT
@PlanBee1
@PleasantLnBees
@pprmedia
@QueenBsHive
@quintassential
@rhodro
@romancingthebee
@sakisbeekeeper
@SelenaGovier
@SheffieldHoney
@solitarybee
@somersetbeeman
@SurreyLife
@TD_Beekeepers
@TheBeeVet
@TheChoirBoy
@TheSoapDragon
@thorncroftclems
@UKHoneyBeeMan
@urban_honey_co
@urbanhoneycoll
@urbfarmbeehives
@val_littlewood
@willipmrpip

'A charming, oddly moving and genuinely useful book'
A. L. Kennedy

Minding my Peas and Cucumbers

Quirky Tales of Allotment Life

Kay Sexton

MINDING MY PEAS AND CUCUMBERS

Quirky Tales of Allotment Life

Kay Sexton

ISBN: 978 1 84953 135 1 Hardback £9.99

When Kay Sexton becomes the proud holder of an allotment, she hopes it will be her first foray towards self-sufficiency for her family. Instead, she finds herself in a strange and hostile world of hosepipe stand-offs and arcane rules and regulations.

She finds her mud-caked, wellingtoned feet and successfully navigates her way through allotment-keeping, battling biblical-scale pest invasions, learning the dark arts of the competitive vegetable grower and practising ninja-like disappearing acts to avoid yet another free cucumber from a neighbouring gardener.

Witty, well-observed and with mouth-watering recipes, this book is for anyone who dreams of a slice of the good life.

'I've dreamed of having an allotment, on and off, for the last 40 years, but the "too busy" excuse has always prevailed. If Kay Sexton's witty, practical and perceptive "Tales" are the closest I'll ever get to fulfilling my allotment dream, then it won't have been all bad' Jonathon Porritt

'Must read' YOURS Magazine

'Guide to the good life' THE FIELD

TRUGS
DIBBERS
TROWELS
AND
TWINE and everything
good about GARDENING
with little TIPS and
words of WISDOM and
INSPIRATION on the simplest
of pleasures
ISOBEL CARLSON

TRUGS, DIBBERS, TROWELS AND TWINE

And Everything Good About Gardening with Little Tips and Words of Wisdom and Inspiration on the Simplest of Pleasures

Isobel Carlson

ISBN: 978 1 84953 040 8 Hardback £9.99

'The love of gardening is a seed once sown that never dies'
GERTRUDE JEKYLL

Follow the garden path to horticultural heaven with this satisfying harvest of useful tips and green-fingered wisdom. Learn about:

- Growing your own herbs to make therapeutic delights
- Banning bugs and slugs and attracting in valuable creatures
- Recycling old household items
- Keeping bees and chickens
- Cooking up your produce, even the weeds

The Gardener's Calendar
Pippa Greenwood

THE GARDENER'S CALENDAR

Pippa Greenwood

ISBN: 978 1 84953 135 1 Hardback £9.99

This wonderful and charmingly illustrated reference guide contains specific month-by-month 'to do' lists for ornamental gardens, edible crops and general maintenance, as well as tips on things to look out for, such as pests and how to eliminate them. With diary pages for making your own notes each month, this pocket-sized calendar is a must-have. Whether you're a seasoned gardener or just starting out, this book will become an old and trusted friend.

Pippa Greenwood was a presenter on BBC Two's *Gardeners' World* for over 13 years. She is a regular panellist and bug expert on BBC Radio 4's *Gardeners' Question Time*, has written a number of award-winning books and contributes to *BBC Gardeners' World* magazine.

Have you enjoyed this book?
If so, why not write a review on your favourite website?

If you're interested in finding out more about our books,
follow us on Twitter: **@summersdale**

Thanks very much for buying this Summersdale book.

www.summersdale.com